Eloi Rylan Koios (Ed.)

4355 Memphis

Eloi Rylan Koios (Ed.)

4355 Memphis

Asteroid, Asteroid Belt, Asteroid Family, Solar System, Trojan (Astronomy)

Tract

Publisher:
Tract is a trademark of
International Book Market Service Ltd., 17 Rue Meldrum, Beau Bassin, 1713-01 Mauritius
Email: info@bookmarketservice.com
Website: www.bookmarketservice.com

Published in 2011

Printed in: U.S.A., U.K., Germany. This book was not produced in Mauritius.

ISBN: 978-613-8-88598-6

Contents

Articles

References

Asteroid

A composite image, to scale, of the asteroids that have been imaged at high resolution. As of 2011 they are, from largest to smallest: 4 Vesta, 21 Lutetia, 253 Mathilde, 243 Ida and its moon Dactyl, 433 Eros, 951 Gaspra, 2867 Šteins, 25143 Itokawa.

The largest asteroid in the previous image, Vesta (left), with Ceres (center) and Earth's Moon (right) shown to scale.

Asteroids (from Greek ἀστήρ 'star' and εἶδος 'like, in form') are a class of small Solar System bodies in orbit around the Sun. They have also been called **planetoids**, especially the larger ones. These terms have historically been applied to any astronomical object orbiting the Sun that did not show the disk of a planet and was not observed to have the characteristics of an active comet, but as small objects in the outer Solar System were discovered, their volatile-based surfaces were found to more closely resemble comets, and so were often distinguished from traditional asteroids.[1] Thus the term *asteroid* has come increasingly to refer specifically to the small rocky and metallic bodies of the inner Solar System out to the orbit of Jupiter. They are grouped with the outer bodies—centaurs, Neptune trojans, and trans-Neptunian objects—as **minor planets**, which is the term preferred in astronomical circles.[2] This article will restrict the use of the term 'asteroid' to the minor planets of the inner Solar System.

There are millions of asteroids, many thought to be the shattered remnants of planetesimals, bodies within the young Sun's solar nebula that never grew large enough to become planets.[3] A large majority of known asteroids orbit in the asteroid belt between the orbits of Mars and Jupiter or co-orbital with Jupiter (the Jupiter Trojans). However, other orbital families exist with significant populations, including the near-Earth asteroids. Individual asteroids are classified by their characteristic spectra, with the majority falling into three main groups: C-type, S-type, and M-type. These were named after and are generally identified with carbon-rich, stony, and metallic compositions, respectively.

Naming

A newly discovered asteroid is given a provisional designation (such as 2002 AT_4) consisting of the year of discovery and an alphanumeric code indicating the half-month of discovery and the sequence within that half-month. Once an asteroid's orbit has been confirmed, it is given a number, and later may also be given a name (e.g. 433 Eros). The formal naming convention uses parentheses around the number (e.g. (433) Eros), but dropping the parentheses is quite common. Informally, it is common to drop the number altogether, or to drop it after the first mention when a name is repeated in running text.

Symbols

The first asteroids to be discovered were assigned iconic symbols like the ones traditionally used to designate the planets. By 1855 there were two dozen asteroid symbols, which often occurred in several variants.[4]

Asteroid	Symbol	
Ceres		Ceres' scythe, reversed to double as the letter *C*
2 Pallas	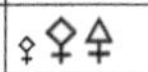	Athena's (Pallas') spear
3 Juno		A star mounted on a scepter, for Juno, the Queen of Heaven
4 Vesta	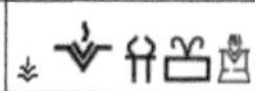	The altar and sacred fire of Vesta
5 Astraea		A scale, or an inverted anchor, symbols of justice
6 Hebe		Hebe's cup
7 Iris		A rainbow (*iris*) and a star
8 Flora		A flower (*flora*) (spec. the Rose of England)
9 Metis		The eye of wisdom and a star
10 Hygiea		Hygiea's serpent and a star, or the Rod of Asclepius
11 Parthenope		A harp, or a fish and a star; symbols of the sirens
12 Victoria		The laurels of victory and a star
13 Egeria		A shield, symbol of Egeria's protection, and a star
14 Irene		A dove carrying an olive-branch (symbol of *irene* 'peace') with a star on its head,[5] or an olive branch, a flag of truce, and a star
15 Eunomia		A heart, symbol of good order (*eunomia*), and a star
16 Psyche		A butterfly's wing, symbol of the soul (*psyche*), and a star
17 Thetis		A dolphin, symbol of Thetis, and a star
18 Melpomene		The dagger of Melpomene, and a star
19 Fortuna		The wheel of fortune and a star
26 Proserpina		Proserpina's pomegranate
28 Bellona		Bellona's whip and lance[6]
29 Amphitrite		The shell of Amphitrite and a star
35 Leukothea		A lighthouse beacon, symbol of Leucothea[7]
37 Fides		The cross of faith (*fides*)[8]

In 1851,[9] Johann Franz Encke made a major change in the upcoming 1854 edition of the *Berliner Astronomisches Jahrbuch* (BAJ, *Berlin Astronomical Yearbook*). He introduced a disk (circle), a traditional symbol for a star, as the generic symbol for an asteroid. The circle was then numbered in order of discovery to indicate a specific asteroid, though he assigned ① to the fifth, Astraea, the first four continuing with their existing symbols. The numbered-circle convention was quickly adopted by the astronomical community, and no iconic symbols were created after 1855.[10] That year Astraea's number was bumped up to ⑤, but Ceres through Vesta would not be listed by their numbers until the 1867 edition. The circle would become a pair of parentheses, and the parentheses sometimes omitted altogether over the next few decades, leading to the modern convention.[5]

Discovery

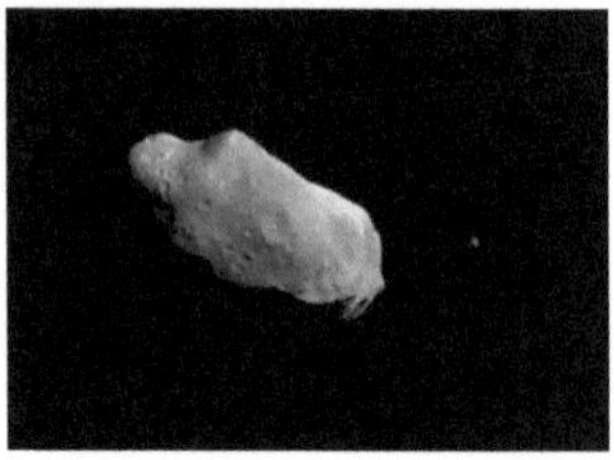
243 Ida and its moon Dactyl. Dactyl is the first satellite of an asteroid to be discovered.

The first asteroid to be discovered, Ceres, was found in 1801 by Giuseppe Piazzi, and was originally considered to be a new planet.[11] This was followed by the discovery of other similar bodies, which with the equipment of the time appeared to be points of light, like stars, showing little or no planetary disc, though readily distinguishable from stars due to their apparent motions. This prompted the astronomer Sir William Herschel to propose the term "asteroid", from Greek *αστεροειδής*, *asteroeidēs* 'star-like, star-shaped', from ancient Greek *αστήρ*, *astēr* 'star, planet'. In the early second half of the nineteenth century, the terms "asteroid" and "planet" (not always qualified as "minor") were still used interchangeably; for example, the *Annual of Scientific Discovery for 1871* [12], page 316, reads "Professor J. Watson has been awarded by the Paris Academy of Sciences, the astronomical prize, Lalande foundation, for the discovery of eight new asteroids in one year. The planet Lydia (No. 110), discovered by M. Borelly at the Marseilles Observatory [...] M. Borelly had previously discovered two planets bearing the numbers 91 and 99 in the system of asteroids revolving between Mars and Jupiter".

Historical methods

Asteroid discovery methods have dramatically improved over the past two centuries.

In the last years of the 18th century, Baron Franz Xaver von Zach organized a group of 24 astronomers to search the sky for the missing planet predicted at about 2.8 AU from the Sun by the Titius-Bode law, partly because of the discovery, by Sir William Herschel in 1781, of the planet Uranus at the distance predicted by the law. This task required that hand-drawn sky charts be prepared for all stars in the zodiacal band down to an agreed-upon limit of faintness. On subsequent nights, the sky would be charted again and any moving object would, hopefully, be spotted. The expected motion of the missing planet was about 30 seconds of arc per hour, readily discernible by observers.

The first object, Ceres, was not discovered by a member of the group, but rather by accident in 1801 by Giuseppe Piazzi, director of the observatory of Palermo in Sicily. He discovered a new star-like object in Taurus and followed the displacement of this object during several nights. His colleague, Carl Friedrich Gauss, used these observations to find the exact distance from this unknown object to the Earth. Gauss' calculations placed the object between the planets Mars and Jupiter. Piazzi named it after Ceres, the Roman goddess of agriculture.

Three other asteroids (2 Pallas, 3 Juno, and 4 Vesta) were discovered over the next few years, with Vesta found in 1807. After eight more years of fruitless searches, most astronomers assumed that there were no more and abandoned any further searches.

However, Karl Ludwig Hencke persisted, and began searching for more asteroids in 1830. Fifteen years later, he found 5 Astraea, the first new asteroid in 38 years. He also found 6 Hebe less than two years later. After this, other astronomers joined in the search and at least one new asteroid was discovered every year after that (except the wartime year 1945). Notable asteroid hunters of this early era were J. R. Hind, Annibale de Gasparis, Robert Luther, H. M. S. Goldschmidt, Jean Chacornac, James Ferguson, Norman Robert Pogson, E. W. Tempel, J. C. Watson, C. H.

F. Peters, A. Borrelly, J. Palisa, the Henry brothers and Auguste Charlois.

In 1891, however, Max Wolf pioneered the use of astrophotography to detect asteroids, which appeared as short streaks on long-exposure photographic plates. This dramatically increased the rate of detection compared with earlier visual methods: Wolf alone discovered 248 asteroids, beginning with 323 Brucia, whereas only slightly more than 300 had been discovered up to that point. It was known that there were many more, but most astronomers did not bother with them, calling them "vermin of the skies", a phrase due to Edmund Weiss.[13] Even a century later, only a few thousand asteroids were identified, numbered and named.

Manual methods of the 1900s and modern reporting

Until 1998, asteroids were discovered by a four-step process. First, a region of the sky was photographed by a wide-field telescope, or Astrograph. Pairs of photographs were taken, typically one hour apart. Multiple pairs could be taken over a series of days. Second, the two films of the same region were viewed under a stereoscope. Any body in orbit around the Sun would move slightly between the pair of films. Under the stereoscope, the image of the body would seem to float slightly above the background of stars. Third, once a moving body was identified, its location would be measured precisely using a digitizing microscope. The location would be measured relative to known star locations.[14]

These first three steps do not constitute asteroid discovery: the observer has only found an apparition, which gets a provisional designation, made up of the year of discovery, a letter representing the half-month of discovery, and finally a letter and a number indicating the discovery's sequential number (example: 1998 FJ_{74}).

The last step of discovery is to send the locations and time of observations to the Minor Planet Center, where computer programs determine whether an apparition ties together earlier apparitions into a single orbit. If so, the object receives a catalogue number and the observer of the first apparition with a calculated orbit is declared the discoverer, and granted the honor of naming the object subject to the approval of the International Astronomical Union.

Computerized methods

There is increasing interest in identifying asteroids whose orbits cross Earth's, and that could, given enough time, collide with Earth (see Earth-crosser asteroids). The three most important groups of near-Earth asteroids are the Apollos, Amors, and Atens. Various asteroid deflection strategies have been proposed, as early as the 1960s.

2004 FH is the center dot being followed by the sequence; the object that flashes by during the clip is an artificial satellite.

The near-Earth asteroid 433 Eros had been discovered as long ago as 1898, and the 1930s brought a flurry of similar objects. In order of discovery, these were: 1221 Amor, 1862 Apollo, 2101 Adonis, and finally 69230 Hermes, which approached within 0.005 AU of the Earth in 1937. Astronomers began to realize the possibilities of Earth impact.

Two events in later decades increased the alarm: the increasing acceptance of Walter Alvarez' hypothesis that an impact event resulted in the Cretaceous-Tertiary extinction, and the 1994 observation of Comet Shoemaker-Levy 9 crashing into Jupiter. The U.S. military also declassified the information that its military satellites, built to detect nuclear explosions, had detected hundreds of upper-atmosphere impacts by objects ranging from one to 10 metres across.

All these considerations helped spur the launch of highly efficient automated systems that consist of Charge-Coupled Device (CCD) cameras and computers directly connected to telescopes. Since 1998, a large majority of the asteroids have been discovered by such automated systems. A list of teams using such automated systems includes:[15]

- The Lincoln Near-Earth Asteroid Research (LINEAR) team
- The Near-Earth Asteroid Tracking (NEAT) team
- Spacewatch
- The Lowell Observatory Near-Earth-Object Search (LONEOS) team
- The Catalina Sky Survey (CSS)
- The Campo Imperatore Near-Earth Objects Survey (CINEOS) team
- The Japanese Spaceguard Association
- The Asiago-DLR Asteroid Survey (ADAS)

The LINEAR system alone has discovered 121,346 asteroids, as of March, 2011.[16] Among all the automated systems, 4711 near-Earth asteroids have been discovered[17] including over 600 more than 1 km (0.6 mi) in diameter.

Terminology

Traditionally, small bodies orbiting the Sun were classified as asteroids, comets or meteoroids, with anything smaller than ten metres across being called a meteoroid.[18] The term "asteroid" is ill-defined. It never had a formal definition, with the broader term minor planet being preferred by the International Astronomical Union from 1853 on. In 2006, the term "small Solar System body" was introduced to cover both most minor planets and comets.[19] Other languages prefer "planetoid" (Greek for "planet-like"), and this term is occasionally used in English for the larger asteroids. The word "planetesimal" has a similar meaning, but refers specifically to the small building blocks of the planets that existed when the Solar System was forming. The term "planetule" was coined by the geologist William Daniel Conybeare to describe minor planets,[20] but is not in common use. The three largest objects in the asteroid belt, Ceres, 2 Pallas, and 4 Vesta, grew to the stage of protoplanets. Ceres has been classified as a dwarf planet, the only one in the inner Solar System.

When found, asteroids were seen as a class of objects distinct from comets, and there was no unified term for the two until "small Solar System body" was coined in 2006. The main difference between an asteroid and a comet is that a comet shows a coma due to sublimation of near surface ices by solar radiation. A few objects have ended up being dual-listed because they were first classified as minor planets but later showed evidence of cometary activity. Conversely, some (perhaps all) comets are eventually depleted of their surface volatile ices and become asteroids. A further distinction is that comets typically have more eccentric orbits than most asteroids; most "asteroids" with notably eccentric orbits are probably dormant or extinct comets.[21]

For almost two centuries, from the discovery of Ceres in 1801 until the discovery of the first centaur, 2060 Chiron, in 1977, all known asteroids spent most of their time at or within the orbit of Jupiter, though a few such as 944 Hidalgo ventured far beyond Jupiter for part of their orbit. When astronomers started finding more small bodies that permanently resided further out than Jupiter, now called centaurs, they numbered them among the traditional asteroids, though there was debate over whether they should be classified as asteroids or as a new type of object. Then, when the first trans-Neptunian object, 1992 QB1, was discovered in 1992, and especially when large numbers of similar objects started turning up, new terms were invented to sidestep the issue: Kuiper-belt object, trans-Neptunian object, scattered-disc object, and so on. These inhabit the cold outer reaches of the Solar System where ices remain solid and comet-like bodies are not expected to exhibit much cometary activity; if centaurs or trans-Neptunian objects were to venture close to the Sun, their volatile ices would sublimate, and traditional approaches would classify them as comets and not asteroids.

The innermost of these are the Kuiper-belt objects, called "objects" partly to avoid the need to classify them as asteroids or comets.[22] They are believed to be predominantly comet-like in composition, though some may be more akin to asteroids.[23] Furthermore, most do not have the highly eccentric orbits associated with comets, and the ones so far discovered are larger than traditional comet nuclei. (The much more distant Oort cloud is hypothesized to be the main reservoir of dormant comets.) Other recent observations, such as the analysis of the cometary dust collected by the Stardust probe, are increasingly blurring the distinction between comets and asteroids,[24] suggesting "a continuum between asteroids and comets" rather than a sharp dividing line.[25]

The minor planets beyond Jupiter's orbit are sometimes also called "asteroids", especially in popular presentations.[26] However, it is becoming increasingly common for the term "asteroid" to be restricted to minor planets of the inner Solar System.[22] Therefore, this article will restrict itself for the most part to the classical asteroids: objects of the asteroid belt, Jupiter trojans, and near-Earth objects.

When the IAU introduced the class small Solar System bodies in 2006 to include most objects previously classified as minor planets and comets, they created the class of dwarf planets for the largest minor planets—those that have enough mass to have become ellipsoidal under their own gravity. According to the IAU, "the term 'minor planet' may still be used, but generally the term 'Small Solar System Body' will be preferred."[27] Currently only the largest object in the asteroid belt, Ceres, at about 950 km (590 mi) across, has been placed in the dwarf planet category, although there are several large asteroids (Vesta, Pallas, and Hygiea) that may be classified as dwarf planets when their shapes are better known.[28]

Formation

It is believed that planetesimals in the asteroid belt evolved much like the rest of the solar nebula until Jupiter neared its current mass, at which point excitation from orbital resonances with Jupiter ejected over 99% of planetesimals in the belt. Simulations and a discontinuity in spin rate and spectral properties suggest that asteroids larger than approximately 120 km (75 mi) in diameter accreted during that early era, whereas smaller bodies are fragments from collisions between asteroids during or after the Jovian disruption.[29] Ceres and Vesta grew large enough to melt and differentiate, with heavy metallic elements sinking to the core, leaving rocky minerals in the crust.[30]

In the Nice model, many Kuiper-belt objects are captured in the outer asteroid belt, at distances greater than 2.6 AU. Most were later ejected by Jupiter, but those that remained may be the D-type asteroids, and possibly include Ceres.[31]

Distribution within the Solar System

Various dynamical groups of asteroids have been discovered orbiting in the inner Solar System. Their orbits are perturbed by the gravity of other bodies in the Solar System and by the Yarkovsky effect. Significant populations include:

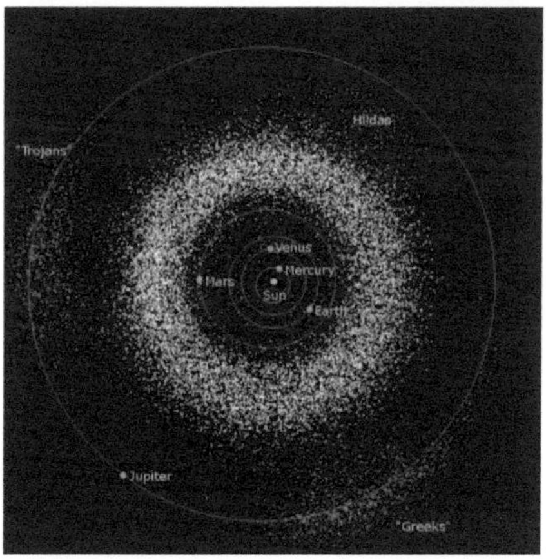

The asteroid belt (white) and the Trojan asteroids (green)

Asteroid belt

The majority of known asteroids orbit within the asteroid belt between the orbits of Mars and Jupiter, generally in relatively low-eccentricity (i.e., not very elongated) orbits. This belt is now estimated to contain between 1.1 and 1.9 million asteroids larger than 1 km (0.6 mi) in diameter,[32] and millions of smaller ones.[33] These asteroids may be remnants of the protoplanetary disk, and in this region the accretion of planetesimals into planets during the formative period of the Solar System was prevented by large gravitational perturbations by Jupiter.

Trojans

Trojan asteroids are a population that share an orbit with a larger planet or moon, but do not collide with it because they orbit in one of the two Lagrangian points of stability, L4 and L5, which lie 60° ahead of and behind the larger body.

The most significant population of Trojan asteroids are the Jupiter Trojans. Although fewer Jupiter Trojans have been discovered as of 2010, it is thought that they are as numerous as the asteroids in the asteroid belt.

A couple trojans have also been found orbiting with Mars.[34]

Near-Earth asteroids

Near-Earth asteroids, or NEAs, are asteroids that have orbits that pass close to that of Earth. Asteroids that actually cross the Earth's orbital path are known as *Earth-crossers*. As of May 2010, 7,075 near-Earth asteroids are known and the number over one kilometre in diameter is estimated to be 500–1,000.

Characteristics

Size distribution

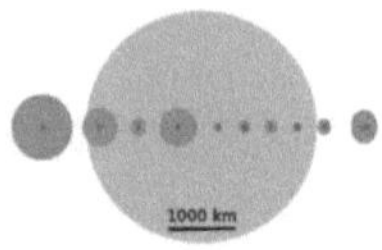

Sizes of the first ten asteroids to be discovered, compared to the Earth's Moon

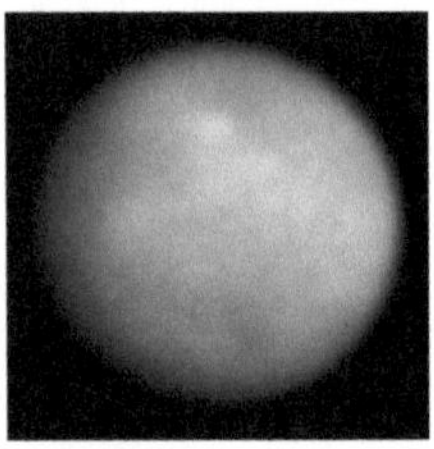
HST image of the dwarf planet Ceres

Asteroids vary greatly in size, from almost 1000 kilometres for the largest down to rocks just tens of metres across.[35] The three largest are very much like miniature planets: they are roughly spherical, have at least partly differentiated interiors,[36] and are thought to be surviving protoplanets. The vast majority, however, are much smaller and are irregularly shaped; they are thought to be either surviving planetesimals or fragments of larger bodies.

The dwarf planet Ceres is by far the largest asteroid, with a diameter of 975 km (610 mi). The next largest are 2 Pallas and 4 Vesta, both with diameters of just over 500 km (300 mi). Vesta is the only main-belt asteroid that can, on occasion, be visible to the naked eye. On some rare occasions, a near-Earth asteroid may briefly become visible without technical aid; see 99942 Apophis.

The mass of all the objects of the asteroid belt, lying between the orbits of Mars and Jupiter, is estimated to be about 2.8-3.2×10^{21} kg, or about 4 percent of the mass of the Moon. Of this, Ceres comprises 0.95×10^{21} kg, a third of the total.[37] Adding in the next three most massive objects, Vesta (9%), Pallas (7%), and Hygiea (3%), brings this figure up to 51%; while the three after that, 511 Davida (1.2%), 704 Interamnia (1.0%), and 52 Europa (0.9%), only add another 3% to the total mass. The number of asteroids then increases rapidly as their individual masses decrease.

The number of asteroids decreases markedly with size. Although this generally follows a power law, there are 'bumps' at 5 km and 100 km, where more asteroids than expected from a logarithmic distribution are found.[38]

Approximate number of asteroids N larger than diameter D

D	100 m	300 m	500 m	1 km	3 km	5 km	10 km	30 km	50 km	100 km	200 km	300 km	500 km	900 km
N	~25,000,000	4,000,000	2,000,000	750,000	200,000	90,000	10,000	1,100	600	200	30	5	3	1

Largest asteroids

Although their location in the asteroid belt excludes them from planet status, the four largest objects, Ceres, Vesta, Pallas, and Hygiea, are remnant protoplanets that share many characteristics common to planets, and are atypical compared to the majority of "potato"-shaped asteroids.

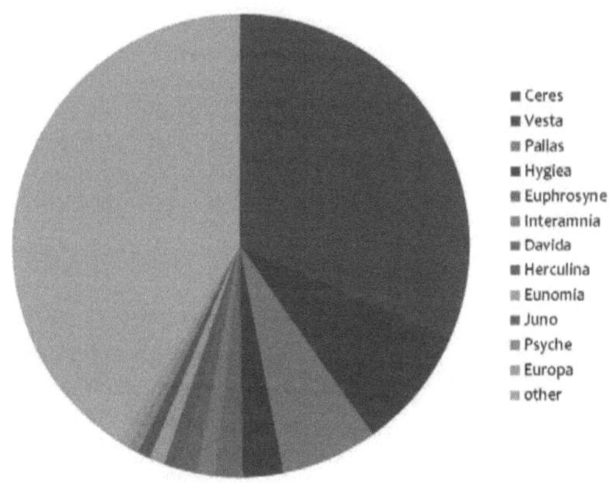

The relative masses of the twelve List of notable asteroids#Largest by masslargest asteroids known, "Recent Asteroid Mass Determinations". Maintained by Jim Baer. Last updated 2010-12-12. Access date 2011-09-02. The values of Juno and Herculina may be off by as much as 16%, and Euphrosyne by a third. The order of the lower eight may change as better data is acquired, but the values do not overlap with any known asteroid outside these twelve. compared to the remaining mass of the asteroid belt. Elena V. PitjevaPitjeva, E. V. (2005). "High-Precision Ephemerides of Planets—EPM and Determination of Some Astronomical Constants" (PDF). Solar System Research 39 (3): 184. Bibcode 2005SoSyR..39..176P. doi:10.1007/s11208-005-0033-2. .

Attributes of protoplanetary asteroids

Name	Orbital radius (AU)	Orbital period (years)	Inclination to ecliptic	Orbital eccentricity	Diameter (km)	Diameter (% of Moon)	Mass ($\times 10^{18}$ kg)	Mass (% of Ceres)	Rotation period (hr)	Axial tilt	Surface temperature
Vesta	2.36	3.63	7.1°	0.089	578×560×458 (mean 529)	15%	260	28%	5.34	29°	85–270 K
Ceres	2.77	4.60	10.6°	0.079	975×975×909 (mean)	28%	940	100%	9.07	≈ 3°	167 K
Pallas	2.77	4.62	34.8°	0.231	580×555×500 (mean 545)	16%	210	22%	7.81	≈ 80°	164 K
Hygiea	3.14	5.56	3.8°	0.117	530×407×370 (mean 430)	12%	87	9%	27.6	≈ 60°	164 K

Ceres is the only asteroid large enough for its gravity to force it into a spheroidal shape, and so, according to the IAU's 2006 resolution on the definition of a planet, it has been classified as a dwarf planet.[41] Vesta may eventually be so classified as well. Ceres has a much higher absolute magnitude than the other asteroids, of around 3.32,[42] and may possess a surface layer of ice.[43] Like the planets, Ceres is differentiated: it has a crust, a mantle and a core.[43] Vesta, too, has a differentiated interior, though it formed inside the Solar System's frost line, and so is devoid of water;[44] its composition is mainly of basaltic rock such as olivine.[] Pallas is unusual in that, like Uranus, it rotates on its side, with one pole regularly facing the Sun and the other facing away.[45] Its composition is similar to that of Ceres: high in carbon and silicon, and perhaps partially differentiated.[46] Hygiea is a carbonaceous asteroid and, unlike the other largest asteroids, lies relatively close to the plane of the ecliptic.[47]

Rotation

Measurements of the rotation rates of large asteroids in the asteroid belt show that there is an upper limit. No asteroid with a diameter larger than 100 meters has a rotation period smaller than 2.2 hours. For asteroids rotating faster than approximately this rate, the inertia at the surface is greater than the gravitational force, so any loose surface material would be flung out. However, a solid object should be able to rotate much more rapidly. This suggests that most asteroids with a diameter over 100 meters are rubble piles formed through accumulation of debris after collisions between asteroids.[48]

Composition

The physical composition of asteroids is varied and in most cases poorly understood. Ceres appears to be composed of a rocky core covered by an icy mantle, where Vesta is thought to have a nickel-iron core, olivine mantle, and basaltic crust.[49] 10 Hygiea, however, which appears to have a uniformly primitive composition of carbonaceous chondrite, is thought to be the largest undifferentiated asteroid. Most of the smaller asteroids are thought to be piles of rubble held together loosely by gravity, though the largest are probably solid. Some asteroids have moons or are co-orbiting binaries: Rubble piles, moons, binaries, and scattered asteroid families are believed to be the results of collisions that disrupted a parent asteroid.

Asteroids contain traces of amino-acids and other organic compounds, and some speculate that asteroid impacts may have seeded the early Earth with the chemicals necessary to initiate life, or may have even brought life itself to Earth. (See also panspermia.)[50] In August 2011, a report, based on NASA studies with meteorites found on Earth, was published suggesting DNA and RNA components (adenine, guanine and related organic molecules) may have been formed on asteroids and comets in outer space.[51] [52] [53]

Only one asteroid, 4 Vesta, which has a reflective surface, is normally visible to the naked eye, and this only in very dark skies when it is favorably positioned. Rarely, small asteroids passing close to Earth may be naked-eye visible for a short time.[54]

Composition is calculated from three primary sources: albedo, surface spectrum, and density. The last can only be determined accurately by observing the orbits of moons the asteroid might have. So far, every asteroid with moons has turned out to be a rubble pile, a loose conglomeration of rock and metal that may be half empty space by volume. The investigated asteroids are as large as 280 km in diameter, and include 121 Hermione (268×186×183 km), and 87 Sylvia (384×262×232 km). Only half a dozen asteroids are larger than 87 Sylvia, though none of them have moons; however, some smaller asteroids are thought to be more massive, suggesting they may not have been disrupted, and indeed 511 Davida, the same size as Sylvia to within measurement error, is estimated to be two and a half times as massive, though this is highly uncertain. The fact that such large asteroids as Sylvia can be rubble piles, presumably due to disruptive impacts, has important consequences for the formation of the Solar system: Computer simulations of collisions involving solid bodies show them destroying each other as often as merging, but colliding rubble piles are more likely to merge. This means that the cores of the planets could have formed relatively quickly.[55]

Surface features

253 Mathilde, a C-type asteroid measuring about 50 kilometres (30 mi) across, covered in craters half that size. Photograph taken in 1997 by the NEAR Shoemaker probe.

Most asteroids outside the big four (Ceres, Pallas, Vesta, and Hygiea) are likely to be broadly similar in appearance, if irregular in shape. 50-km 253 Mathilde (shown at right) is a rubble pile saturated with craters with diameters the size of the asteroid's radius, and Earth-based observations of 300-km 511 Davida, one of the largest asteroids after the big four, reveal a similarly angular profile, suggesting it is also saturated with radius-size craters.[56] Medium-sized asteroids such as Mathilde and 243 Ida that have been observed up close also reveal a deep regolith covering the surface. Of the big four, Pallas and Hygiea are practically unknown. Vesta has compression fractures encircling a radius-size crater at its south pole but is otherwise a spheroid. Ceres seems quite different in the glimpses Hubble has provided, with surface features that are unlikely to be due to simple craters and impact basins, but details will not be known until *Dawn* arrives in 2015.

Classification

Asteroids are commonly classified according to two criteria: the characteristics of their orbits, and features of their reflectance spectrum.

Orbital classification

Many asteroids have been placed in groups and families based on their orbital characteristics. Apart from the broadest divisions, it is customary to name a group of asteroids after the first member of that group to be discovered. Groups are relatively loose dynamical associations, whereas families are tighter and result from the catastrophic break-up of a large parent asteroid sometime in the past.[57] Families have only been recognized within the asteroid belt. They were first recognised by Kiyotsugu Hirayama in 1918 and are often called Hirayama families in his honor.

About 30% to 35% of the bodies in the asteroid belt belong to dynamical families each thought to have a common origin in a past collision between asteroids. A family has also been associated with the plutoid dwarf planet Haumea.

Quasi-satellites and horseshoe objects

Some asteroids have unusual horseshoe orbits that are co-orbital with the Earth or some other planet. Examples are 3753 Cruithne and 2002 AA_{29}. The first instance of this type of orbital arrangement was discovered between Saturn's moons Epimetheus and Janus.

Sometimes these horseshoe objects temporarily become quasi-satellites for a few decades or a few hundred years, before returning to their earlier status. Both Earth and Venus are known to have quasi-satellites.

Such objects, if associated with Earth or Venus or even hypothetically Mercury, are a special class of Aten asteroids. However, such objects could be associated with outer planets as well.

Spectral classification

This picture of 433 Eros shows the view looking from one end of the asteroid across the gouge on its underside and toward the opposite end. Features as small as 35 m (115 ft) across can be seen.

In 1975, an asteroid taxonomic system based on colour, albedo, and spectral shape was developed by Clark R. Chapman, David Morrison, and Ben Zellner.[58] These properties are thought to correspond to the composition of the asteroid's surface material. The original classification system had three categories: C-types for dark carbonaceous objects (75% of known asteroids), S-types for stony (silicaceous) objects (17% of known asteroids) and U for those that did not fit into either C or S. This classification has since been expanded to include many other asteroid types. The number of types continues to grow as more asteroids are studied.

The two most widely used taxonomies now used are the Tholen classification and SMASS classification. The former was proposed in 1984 by David J. Tholen, and was based on data collected from an eight-color asteroid survey performed in the 1980s. This resulted in 14 asteroid categories.[59] In 2002, the Small Main-Belt Asteroid Spectroscopic Survey resulted in a modified version of the Tholen taxonomy with 24 different types. Both systems have three broad categories of C, S, and X asteroids, where X consists of mostly metallic asteroids, such as the M-type. There are also several smaller classes.[60]

Note that the proportion of known asteroids falling into the various spectral types does not necessarily reflect the proportion of all asteroids that are of that type; some types are easier to detect than others, biasing the totals.

Problems

Originally, spectral designations were based on inferences of an asteroid's composition.[61] However, the correspondence between spectral class and composition is not always very good, and a variety of classifications is in use. This has led to significant confusion. While asteroids of different spectral classifications are likely to be composed of different materials, there are no assurances that asteroids within the same taxonomic class are composed of similar materials.

At present, the spectral classification based on several coarse resolution spectroscopic surveys in the 1990s is still the standard. Scientists cannot agree on a better taxonomic system, largely due to the difficulty of obtaining detailed measurements consistently for a large sample of asteroids (e.g. finer resolution spectra, or non-spectral data such as densities would be very useful).

Exploration

Until the age of space travel, objects in the asteroid belt were merely pinpricks of light in even the largest telescopes and their shapes and terrain remained a mystery. The best modern ground-based telescopes and the Earth-orbiting Hubble Space Telescope can resolve a small amount of detail on the surfaces of the largest asteroids, but even these mostly remain little more than fuzzy blobs. Limited information about the shapes and compositions of asteroids can be inferred from their light curves (their variation in brightness as they rotate) and their spectral properties, and asteroid sizes can be estimated by timing the lengths of star occulations (when an asteroid passes directly in front of a star). Radar imaging can yield good information about asteroid shapes and orbital and rotational parameters, especially for near-Earth asteroids. In terms of delta v and propellant requirements, NEOs are more easily accessible than the Moon.[62]

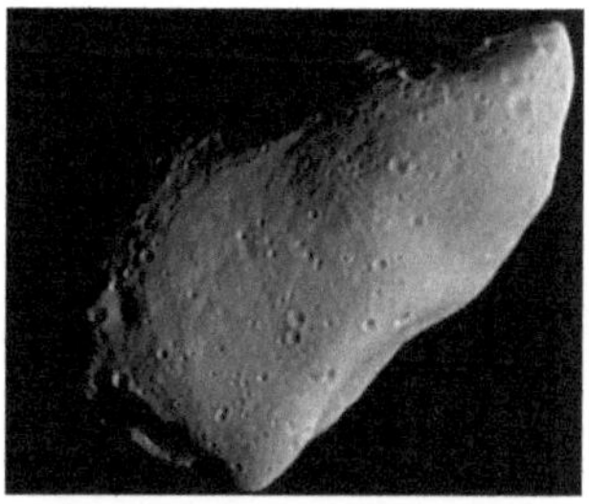

951 Gaspra is the first asteroid to be imaged in close-up.

Vesta, imaged by the Dawn spacecraft

The first close-up photographs of asteroid-like objects were taken in 1971 when the Mariner 9 probe imaged Phobos and Deimos, the two small moons of Mars, which are probably captured asteroids. These images revealed the irregular, potato-like shapes of most asteroids, as did later images from the Voyager probes of the small moons of the gas giants.

The first true asteroid to be photographed in close-up was 951 Gaspra in 1991, followed in 1993 by 243 Ida and its moon Dactyl, all of which were imaged by the Galileo probe en route to Jupiter.

The first dedicated asteroid probe was NEAR Shoemaker, which photographed 253 Mathilde in 1997, before entering into orbit around 433 Eros, finally landing on its surface in 2001.

Other asteroids briefly visited by spacecraft en route to other destinations include 9969 Braille (by Deep Space 1 in 1999), and 5535 Annefrank (by Stardust in 2002).

In September 2005, the Japanese Hayabusa probe started studying 25143 Itokawa in detail and was plagued with difficulties, but returned samples of its surface to earth on June 13, 2010.

The European Rosetta probe (launched in 2004) flew by 2867 Šteins in 2008 and 21 Lutetia, the second-largest asteroid visited to date, in 2010.

In September 2007, NASA launched the Dawn Mission, which started orbiting the protoplanet 4 Vesta in July 2011, and is to orbit 1 Ceres in 2015. 4 Vesta is the largest asteroid visited to date.

In May 2011, NASA announced the OSIRIS-REx sample return mission to asteroid 1999 RQ36, and is expected to launch in 2016.

It has been suggested that asteroids might be used as a source of materials that may be rare or exhausted on earth (asteroid mining), or materials for constructing space habitats (see Colonization of the asteroids). Materials that are heavy and expensive to launch from earth may someday be mined from asteroids and used for space manufacturing and construction.

Fiction

Asteroids and the asteroid belt are a staple of science fiction stories. Asteroids play several potential roles in science fiction: as places human beings might colonize, resources for extracting minerals, hazards encountered by spaceships traveling between two other points, and as a threat to life on Earth by potential impact.

See also

- Asteroid belt
- Asteroid mining
- Mission Marco Polo
- Rosetta probe
- BOOTES (Burst Observer and Optical Transient Exploring System)
- Category:Asteroid groups and families
- Category:Asteroids
- Category:Binary asteroids
- Centaur (minor planet)
- Dwarf planet
- Impact event
- Asteroid deflection strategies
- List of asteroids named after people
- List of asteroids named after places
- List of minor planets
- List of notable asteroids
- Meanings of asteroid names
- Mesoplanet
- Minor planet
- Minor Planet Center
- Near-Earth object
- Orion asteroid mission
- Pioneer 10 space probe
- Pronunciation of asteroid names

Notes

[1] "Asteroids" (http://ssd.jpl.nasa.gov/?asteroids). NASA – Jet Propulsion Laboratory. . Retrieved 13 September2010.

[2] Asimov, Isaac, and Dole, Stephen H. *Planets for Man* (New York: Random House, 1964), p.43

[3] "What Are Asteroids And Comets?" (http://neo.jpl.nasa.gov/faq/#ast). *Near Earth Object Program FAQ*. NASA. . Retrieved 2010-09-13.

[4] Gould, B. A. (1852). "On the Symbolic Notation of the Asteroids". *Astronomical Journal* **2**: 80. Bibcode 1852AJ......2...80G. doi:10.1086/100212.

[5] Hilton, James L. (2001-09-17). "When Did the Asteroids Become Minor Planets" (http://aa.usno.navy.mil/faq/docs/minorplanets.php). . Retrieved 2006-03-26.

[6] Encke, J. F. (1854). "Beobachtung der Bellona, nebst Nachrichten über die Bilker Sternwarte". *Astronomische Nachrichten* **38** (9): 143. doi:10.1002/asna.18540380907.

[7] Rümker, G. (1855). "Name und Zeichen des von Herrn R. Luther zu Bilk am 19. April entdeckten Planeten". *Astronomische Nachrichten* **40** (24): 373. doi:10.1002/asna.18550402405.

[8] Luther, R. (1856). "Schreiben des Herrn Dr. R. Luther, Directors der Sternwarte zu Bilk, an den Herausgeber". *Astronomische Nachrichten* **42** (7): 107. Bibcode 1855AN.....42..107L. doi:10.1002/asna.18550420705.

[9] "When did the asteroids become minor planets?" (http://www.usno.navy.mil/USNO/astronomical-applications/astronomical-information-center/minor-planets). Naval Meteorology and Oceanography Command. . Retrieved 2011-11-06.

[10] Except for Pluto and, in the astrological community, for a few outer bodies such as 2060 Chiron: ⚷

[11] Ceres is the largest asteroid and is now classified as a dwarf planet. All other asteroids are now classified as small Solar System bodies along with comets, centaurs, and the smaller trans-Neptunian objects.

[12] http://books.google.ca/books?id=NAMAAAAAMAAJ&pg=PA316&dq=%22planets%22+asteroids

[13] Seares, Frederick H. (1930). "Address of the Retiring President of the Society in Awarding the Bruce Medal to Professor Max Wolf". *Publ. Astr. Soc. Pacific* **42**: 5–22. Bibcode 1930PASP...42....5S. doi:10.1086/123986+(Dead+links).

[14] Chapman, Mary G. (May 17, 1992). "Carolyn Shoemaker, Planetary Astronomer and Most Successful 'Comet Hunter' To Date" (http://astrogeology.usgs.gov/About/People/CarolynShoemaker). USGS. . Retrieved 2008-04-15.

[15] Yeomans, Don. "Near Earth Object Search Programs" (http://neo.jpl.nasa.gov/programs/). NASA. . Retrieved 2008-04-15.

[16] "Minor Planet Discover Sites" (http://www.minorplanetcenter.org/iau/lists/MPDiscSites.html). . Retrieved 2010-08-24.

[17] "Unusual Minor Planets" (http://www.minorplanetcenter.org/iau/lists/Unusual.html). . Retrieved 2010-08-24.

[18] Beech, M. (September 1995). "On the Definition of the Term Meteoroid". *Quarterly Journal of the Royal Astronomical Society* **36** (3): 281–284. Bibcode 1995QJRAS..36..281B.

[19] The definition of "small Solar System bodies" says that they "include most of the Solar System asteroids, most trans-Neptunian objects, comets, and other small bodies". *The Final IAU Resolution on the definition of "planet" ready for voting* (http://www.iau2006.org/mirror/www.iau.org/iau0602/index.html) (IAU)

[20] "English Dictionary – Browsing Page P-44" (http://www.hyperdictionary.com/dict-e/p-44.html). HyperDictionary.com. . Retrieved 2008-04-15.

[21] Weissman, Paul R., William F. Bottke, Jr., and Harold F. Levinson. "Evolution of Comets into Asteroids." *Southwest Research Institute, Planetary Science Directorate.* 2002. Web Retrieved 3 Aug. 2010 (http://www.boulder.swri.edu/~hal/PDF/asteroids3.pdf)

[22] "Are Kuiper Belt Objects asteroids?" (http://curious.astro.cornell.edu/question.php?number=601), "Ask an astronomer", Cornell University

[23] "Asteroids and Comets" (http://rst.gsfc.nasa.gov/Sect19/Sect19_22.html), NASA website

[24] "Comet Dust Seems More Asteroidy" (http://www.sciam.com/podcast/episode.cfm?id=ADD0878B-D6C3-3B70-7B5BC373545BB82D) *Scientific American*, January 25, 2008

[25] "Comet samples are surprisingly asteroid-like" (http://space.newscientist.com/channel/solar-system/comets-asteroids/dn13224-comet-samples-are-surprisingly-asteroidlike.html), *New Scientist*, 24 January 2008

[26] For instance, a joint NASA-JPL public-outreach website states:

> "We include Trojans (bodies captured in Jupiter's 4th and 5th Lagrange points), Centaurs (bodies in orbit between Jupiter and Neptune), and trans-Neptunian objects (orbiting beyond Neptune) in our definition of "asteroid" as used on this site, even though they may more correctly be called "minor planets" instead of asteroids."

<http://ssd.jpl.nasa.gov/?asteroids>

[27] Questions and Answers on Planets (http://www.iau.org/public_press/news/release/iau0603/questions_answers/), IAU

[28] "Three new planets may join solar system" (http://space.newscientist.com/article.ns?id=dn9761&feedId=online-news_rss20F53), *New Scientist*, 16 August 2006

[29] Bottke, Durda; Nesvorny, Jedicke; Morbidelli, Vokrouhlicky; Levison (2005). "The fossilized size distribution of the main asteroid belt" (http://astro.mff.cuni.cz/davok/papers/fossil05.pdf). *Icarus* **175**: 111. Bibcode 2005Icar..175..111B. doi:10.1016/j.icarus.2004.10.026. .

[30] Kerrod, Robin (2000). *Asteroids, Comets, and Meteors*. Lerner Publications Co.. ISBN 0585317631.

[31] William B. McKinnon, 2008, "On The Possibility Of Large KBOs Being Injected Into The Outer Asteroid Belt". (http://adsabs.harvard.edu/abs/2008DPS....40.3803M) *American Astronomical Society*, DPS meeting #40, #38.03

[32] Tedesco, Edward; Metcalfe, Leo (April 4, 2002). "New study reveals twice as many asteroids as previously believed" (http://www.spaceref.com/news/viewpr.html?pid=7925) (Press release). European Space Agency. . Retrieved 2008-02-21.

[33] World Book at NASA (http://www.nasa.gov/worldbook/asteroid_worldbook.html)

[34] Neptune also has a few known trojans, and these are thought to actually be much more numerous than the Jovian trojans. However, they are often included in the trans-Neptunian population rather than counted with the asteroids.

[35] Below 10 metres, these rocks are by convention considered to be meteoroids.

[36] Schmidt, B.; Russell, C. T.; Bauer, J. M.; Li, J.; McFadden, L. A.; Mutchler, M.; Parker, J. W.; Rivkin, A. S.; Stern, S. A.; Thomas, P. C. (2007). "Hubble Space Telescope Observations of 2 Pallas". *American Astronomical Society, DPS meeting #39* **39**: 485. Bibcode 2007DPS....39.3519S.

[37] Pitjeva, E. V. (2004). "Estimations of masses of the largest asteroids and the main asteroid belt from ranging to planets, Mars orbiters and landers" (http://adsabs.harvard.edu/abs/2004cosp.meet.2014P). *35th COSPAR Scientific Assembly. Held 18–25 July 2004, in Paris, France*. pp. 2014. .

[38] Davis 2002, "Asteroids III", cited by Željko Ivezić (http://www.astro.washington.edu/users/ivezic/Astr598/lecture4.pdf)

[39] "Recent Asteroid Mass Determinations" (http://home.earthlink.net/~jimbaer1/astmass.txt). Maintained by Jim Baer. Last updated 2010-12-12. Access date 2011-09-02. The values of Juno and Herculina may be off by as much as 16%, and Euphrosyne by a third. The order of the lower eight may change as better data is acquired, but the values do not overlap with any known asteroid outside these twelve.

[40] Pitjeva, E. V. (2005). "High-Precision Ephemerides of Planets—EPM and Determination of Some Astronomical Constants" (http://iau-comm4.jpl.nasa.gov/EPM2004.pdf) (PDF). *Solar System Research* **39** (3): 184. Bibcode 2005SoSyR..39..176P.

doi:10.1007/s11208-005-0033-2. .

[41] "The Final IAU Resolution on the Definition of "Planet" Ready for Voting" (http://www.iau.org/public_press/news/detail/iau0602/). IAU. August 24, 2006. . Retrieved 2007-03-02.

[42] Parker, J. W.; Stern, S. A.; Thomas, P. C.; Festou, M. C.; Merline, W. J.; Young, E. F.; Binzel, R. P.; and Lebofsky, L. A. (2002). "Analysis of the First Disk-resolved Images of Ceres from Ultraviolet Observations with the Hubble Space Telescope". *The Astronomical Journal* **123** (1): 549–557. arXiv:astro-ph/0110258. Bibcode 2002AJ....123..549P. doi:10.1086/338093.

[43] "Asteroid 1 Ceres" (http://www.planetary.org/explore/topics/asteroids_and_comets/ceres.html). *The Planetary Society*. . Retrieved 2007-10-20.

[44] "Key Stages in the Evolution of the Asteroid Vesta" (http://hubblesite.org/newscenter/newsdesk/archive/releases/1995/20/image/c). *Hubble Space Telescope news release*. 1995. . Retrieved 2007-10-20. Russel, C. T.; *et al.* (2007). "Dawn mission and operations" (http://journals.cambridge.org/action/displayAbstract?fromPage=online&aid=414750). *NASA/JPL*. . Retrieved 2007-10-20.

[45] Torppa, J.; *et al.* (1996). "Shapes and rotational properties of thirty asteroids from photometric data". *Icarus* **164** (2): 346–383. Bibcode 2003Icar..164..346T. doi:10.1016/S0019-1035(03)00146-5.

[46] Larson, H. P.; Feierberg, M. A.; and Lebofsky, L. A. (1983). "The composition of asteroid 2 Pallas and its relation to primitive meteorites" (http://adsabs.harvard.edu/abs/1983Icar...56..398L). . Retrieved 2007-10-20.

[47] Barucci, M. A.; *et al.* (2002). "10 Hygiea: ISO Infrared Observations" (http://www.lesia.obspm.fr/~crovisier/biblio/preprint/bar02_icarus.pdf) (PDF). . Retrieved 2007-10-21.

"Ceres the Planet" (http://www.orbitsimulator.com/gravity/articles/ceres.html). *orbitsimulator.com*. . Retrieved 2007-10-20.

[48] Rossi, Alessandro (2004-05-20). "The mysteries of the asteroid rotation day" (http://spaceguard.iasf-roma.inaf.it/tumblingstone/issues/current/eng/ast-day.htm). The Spaceguard Foundation. . Retrieved 2007-04-09.

[49] HubbleSite - NewsCenter - Asteroid or Mini-Planet? Hubble Maps the Ancient Surface of Vesta (04/19/1995) - Release Images (http://hubblesite.org/newscenter/archive/releases/1995/20/image/)

[50] Life is Sweet: Sugar-Packing Asteroids May Have Seeded Life on Earth (http://www.space.com/scienceastronomy/planetearth/meteor_sugar_011219.html), Space.com, 19 December 2001

[51] Callahan, M.P.; Smith, K.E.; Cleaves, H.J.; Ruzica, J.; Stern, J.C.; Glavin, D.P.; House, C.H.; Dworkin, J.P. (11 August 2011). "Carbonaceous meteorites contain a wide range of extraterrestrial nucleobases" (http://www.pnas.org/content/early/2011/08/10/1106493108). PNAS. doi:10.1073/pnas.1106493108. . Retrieved 2011-08-15.

[52] Steigerwald, John (8 August 2011). "NASA Researchers: DNA Building Blocks Can Be Made in Space" (http://www.nasa.gov/topics/solarsystem/features/dna-meteorites.html). NASA. . Retrieved 2011-08-10.

[53] ScienceDaily Staff (9 August 2011). "DNA Building Blocks Can Be Made in Space, NASA Evidence Suggests" (http://www.sciencedaily.com/releases/2011/08/110808220659.htm). ScienceDaily. . Retrieved 2011-08-09.

[54] Closest Flyby of Large Asteroid to be Naked-Eye Visible (http://www.space.com/spacewatch/050204_2004_mn4.html), Space.com, 4 February 2005

[55] Marchis, Descamps, et al. *Icarus*, Feb. 2011

[56] A.R. Conrad *et al.* 2007. " Direct measurement of the size, shape, and pole of 511 Davida with Keck AO in a single night (http://www2.keck.hawaii.edu/inst/people/conrad/research/pub/d511.pdf)", *Icarus,* doi:10.1016/j.icarus.2007.05.004

[57] Zappalà, V. (1995). "Asteroid families: Search of a 12,487-asteroid sample using two different clustering techniques". *Icarus* **116** (2): 291–314. Bibcode 1995Icar..116..291Z. doi:10.1006/icar.1995.1127.

[58] Chapman, C. R. (1975). "Surface properties of asteroids: A synthesis of polarimetry, radiometry, and spectrophotometry". *Icarus* **25** (1): 104–130. Bibcode 1975Icar...25..104C. doi:10.1016/0019-1035(75)90191-8.

[59] Tholen, D. J. (March 8–11, 1988). "Asteroid taxonomic classifications" (http://adsabs.harvard.edu/abs/1989aste.conf.1139T). *Asteroids II; Proceedings of the Conference*. Tucson, AZ: University of Arizona Press. pp. 1139–1150. . Retrieved 2008-04-14.

[60] Bus, S. J. (2002). "Phase II of the Small Main-belt Asteroid Spectroscopy Survey: A feature-based taxonomy". *Icarus* **158** (1): 146. Bibcode 2002Icar..158..146B. doi:10.1006/icar.2002.6856.

[61] McSween Jr., Harry Y. (1999). *Meteorites and their Parent Planets* (2nd ed.). Oxford University Press. ISBN 0521587514.

[62] A Piloted Orion Flight to a Near-Earth Object: A Feasibility Study (http://ti.arc.nasa.gov/m/project/neo/pdf/NEO_feasibility.pdf)

References

External links

- Rocks from the Main Belt asteroids (http://rocksfromspace.open.ac.uk/asteroid_belt_detail.htm)
- Alphabetical list of minor planet names (ASCII) (http://www.minorplanetcenter.org/iau/lists/MPNames.html) (Minor Planet Center)
- Near Earth Asteroid Tracking (NEAT) (http://neat.jpl.nasa.gov/)
- Asteroids Page (http://solarsystem.nasa.gov/planets/profile.cfm?Object=Asteroids) at NASA's Solar System Exploration (http://solarsystem.nasa.gov/)
- Asteroid Simulator with Moon and Earth (http://www.colorado.edu/physics/2000/applets/satellites.html)
- Alphabetical and numerical lists of minor planet names (Unicode) (http://www.ipa.nw.ru/PAGE/DEPFUND/LSBSS/englenam.htm) (Institute of Applied Astronomy)
- Future Asteroid Interception Research (http://www.fair-society.org/)
- Near Earth Objects Dynamic Site (http://newton.dm.unipi.it/cgi-bin/neodys/neoibo)
- Asteroids Dynamic Site (http://hamilton.dm.unipi.it/astdys/) Up-to-date osculating orbital elements and proper orbital elements University of Pisa, Italy.
- JPL small bodies database (http://ssd.jpl.nasa.gov/?sb_elem) Current down-loadable ASCII table of orbit data and absolute mags H for over 200000 asteroids, sorted by number. Caltech/JPL.
- Asteroid naming statistics (http://quasar.ipa.nw.ru/PAGE/DEPFUND/LSBSS/statmpn.htm)
- Spaceguard UK (http://www.spaceguarduk.com/)
- Committee on Small Body Nomenclature (http://www.ss.astro.umd.edu/IAU/csbn/)
- List of minor planet orbital groupings and families from ProjectPluto (http://www.projectpluto.com/mp_group.htm)
- Cunningham, Clifford, "Introduction to Asteroids: The Next Frontier", ISBN 0-943396-16-6
- James L. Hilton: When Did the Asteroids Become Minor Planets? (http://aa.usno.navy.mil/faq/docs/minorplanets.php)
- Kirkwood, Daniel; *Relations between the Motions of some of the Minor Planets* (1874). (http://adsabs.harvard.edu/cgi-bin/nph-bib_query?bibcode=1874MNRAS..35...61K&db_key=AST&high=40daf3f6f901929)
- Schmadel, L.D. (2003). *Dictionary of Minor Planet Names.* 5th ed. IAU/Springer-Verlag: Heidelberg.
- Asteroid articles in Planetary Science Research Discoveries (http://www.psrd.hawaii.edu/Archive/Archive-Asteroids.html)
- Catalogue of the Solar System Small Bodies Orbital Evolution (http://smallbodies.ru/en/)

rue:Астероід

Asteroid_belt

The **asteroid belt** is the region of the Solar System located roughly between the orbits of the planets Mars and Jupiter. It is occupied by numerous irregularly shaped bodies called asteroids or minor planets. The asteroid belt is also termed the **main asteroid belt** or **main belt** because there are other asteroids in the Solar System such as near-Earth asteroids and trojan asteroids. About half the mass of the belt is contained in the four largest asteroids: Ceres, 4 Vesta, 2 Pallas, and 10 Hygiea. These have mean diameters of more than 400 km, while Ceres, the asteroid belt's only identified dwarf planet, is about 950 km in diameter.[1] [2] [3] [4] The remaining bodies range down to the size of a dust particle. The asteroid material is so thinly distributed that multiple unmanned spacecraft have traversed it without incident. Nonetheless, collisions between large asteroids do occur, and these can form an asteroid family whose members have similar orbital characteristics and compositions. Collisions also produce a fine dust that forms a major component of the zodiacal light. Individual asteroids within the asteroid belt are categorized by their spectra, with most falling into three basic groups: carbonaceous (C-type), silicate (S-type), and metal-rich (M-type).

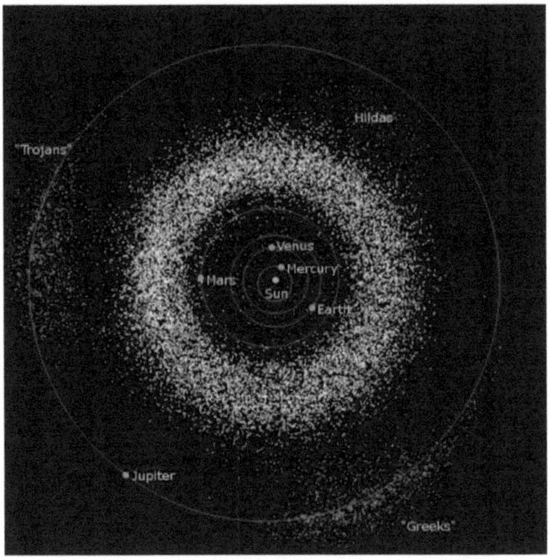

The asteroid belt (shown in white) is located between the orbits of Mars and Jupiter.

The asteroid belt formed from the primordial solar nebula as a group of planetesimals, the smaller precursors of the planets, which in turn formed protoplanets. Between Mars and Jupiter, however, gravitational perturbations from the giant planet imbued the protoplanets with too much orbital energy for them to accrete into a planet. Collisions became too violent, and instead of sticking together, the planetesimals and most of the protoplanets shattered. As a result, most of the asteroid belt's mass has been lost since the formation of the Solar System. Some fragments can eventually find their way into the inner Solar System, leading to meteorite impacts with the inner planets. Asteroid orbits continue to be appreciably perturbed whenever their period of revolution about the Sun forms an orbital resonance with Jupiter. At these orbital distances, a Kirkwood gap occurs as they are swept into other orbits.

Other regions of small Solar System bodies include the centaurs, the Kuiper belt and scattered disk, and the Oort cloud.

History of observation

Giuseppe Piazzi, discoverer of Ceres, known as a planet for many years, then as asteroid number 1, and eventually, a dwarf planet

In an anonymous footnote to his 1766 translation of Charles Bonnet's *Contemplation de la Nature*,[5] the astronomer Johann Daniel Titius of Wittenburg[6] [7] noted an apparent pattern in the layout of the planets. If one began a numerical sequence at 0, then included 3, 6, 12, 24, 48, etc., doubling each time, and added four to each number and divided by 10, this produced a remarkably close approximation to the radii of the orbits of the known planets as measured in astronomical units. This pattern, now known as the Titius–Bode law, predicted the semi-major axes of the six planets of the time (Mercury, Venus, Earth, Mars, Jupiter and Saturn) provided one allowed for a "gap" between the orbits of Mars and Jupiter. In his footnote Titius declared, "But should the Lord Architect have left that space empty? Not at all."[6] In 1768, the astronomer Johann Elert Bode made note of Titius's relationship in his *Anleitung zur Kenntniss des gestirnten Himmels* (English: *Instruction for the Knowledge of the Starry Heavens*) but did not credit Titius, which led many to refer to it as "Bode's law".[7] When William Herschel discovered Uranus in 1781, the planet's orbit matched the law almost perfectly, leading astronomers to conclude that there had to be a planet between the orbits of Mars and Jupiter.

In 1800 the astronomer Baron Franz Xaver von Zach recruited 24 of his fellows into a club, the Vereinigte Astronomische Gesellschaft ("United Astronomical Society") which he informally dubbed the "Lilienthal Society"[8] for its meetings in Lilienthal, a small city near Bremen. Determined to bring the Solar System to order, the group became known as the "Himmelspolizei", or Celestial Police. Notable members included Herschel, the British Astronomer Royal Nevil Maskelyne, Charles Messier, and Heinrich Olbers.[9] The Society assigned to each astronomer a 15° region of the zodiac to search for the missing planet.[10]

Only a few months later, a non-member of the Celestial Police confirmed their expectations. On January 1, 1801, Giuseppe Piazzi, Chair of Astronomy at the University of Palermo, Sicily, found a tiny moving object in an orbit with exactly the radius predicted by the Titius–Bode law. He dubbed it Ceres, after the Roman goddess of the harvest and patron of Sicily. Piazzi initially believed it a comet, but its lack of a coma suggested it was a planet.[9] Fifteen months later, Olbers discovered a second object in the same region, Pallas. Unlike the other known planets, the objects remained points of light even under the highest telescope magnifications, rather than resolving into discs. Apart from their rapid movement, they appeared indistinguishable from stars. Accordingly, in 1802 William Herschel suggested they be placed into a separate category, named *asteroids*, after the Greek *asteroeides*, meaning "star-like".[11][12] Upon completing a series of observations of Ceres and Pallas, he concluded,[13]

> Neither the appellation of planets, nor that of comets, can with any propriety of language be given to these two stars ... They resemble small stars so much as hardly to be distinguished from them. From this, their asteroidal appearance, if I take my name, and call them Asteroids; reserving for myself however the liberty of changing that name, if another, more expressive of their nature, should occur.

Despite Herschel's coinage, for several decades it remained common practice to refer to these objects as planets.[5] By 1807, further investigation revealed two new objects in the region: 3 Juno and 4 Vesta.[14] The burning of Lilienthal in the Napoleonic wars brought this first period of discovery to a close,[14] and only in 1845 did astronomers detect another object (5 Astraea). Shortly thereafter new objects were found at an accelerating rate, and counting them among the planets became increasingly cumbersome. Eventually, they were dropped from the planet list as first suggested by Alexander von Humboldt in the early 1850s, and William Herschel's choice of nomenclature, "asteroids", gradually came into common use.[5]

The discovery of Neptune in 1846 led to the discrediting of the Titius–Bode law in the eyes of scientists, as its orbit was nowhere near the predicted position. To date, there is no scientific explanation for the law, and astronomers' consensus regards it as a coincidence.[15]

The expression "asteroid belt" came into use in the very early 1850s, although it is hard to pinpoint who coined the term. The first English use seems to be in the 1850 translation (by E. C. Otté) of Alexander von Humboldt's *Cosmos:*[16] "[...] and the regular appearance, about the 13th of November and the 11th of August, of shooting stars, which probably form part of a belt of asteroids intersecting the Earth's orbit and moving with planetary velocity". Other early appearances occur in Robert James Mann's *A Guide to the Knowledge of the Heavens*,[17] "The orbits of the asteroids are placed in a wide belt of space, extending between the extremes of [...]". The American astronomer Benjamin Peirce seems to have adopted that terminology and to have been one of its promoters.[18] One hundred asteroids had been located by mid-1868, and in 1891 the introduction of astrophotography by Max Wolf accelerated the rate of discovery still further.[19] A total of 1,000 asteroids had been found by 1921,[20] 10,000 by 1981,[21] and 100,000 by 2000.[22] Modern asteroid survey systems now use automated means to locate new minor planets in ever-increasing quantities.

Origin

Formation

In 1802, shortly after discovering Pallas, Heinrich Olbers suggested to William Herschel that Ceres and Pallas were fragments of a much larger planet that once occupied the Mars-Jupiter region, this planet having suffered an internal explosion or a cometary impact many million years before.[23] Over time, however, this hypothesis has fallen from favor. The large amount of energy that would have been required to destroy a planet, combined with the belt's low combined mass, which is only about 4% of the mass of the Earth's Moon, do not support the hypothesis. Further, the significant chemical differences between the asteroids are difficult to explain if they come from the same planet.[24] Today, most scientists accept that, rather than fragmenting from a progenitor planet, the asteroids never formed a planet at all.

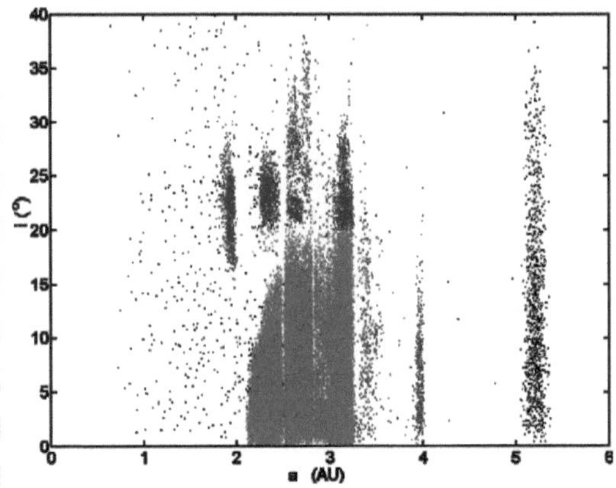

The asteroid belt showing the orbital inclinations versus distances from the Sun, with asteroids in the core region of the asteroid belt in red and other asteroids in blue

In general in the Solar System, planetary formation is thought to have occurred via a process comparable to the long-standing nebular hypothesis: a cloud of interstellar dust and gas collapsed under the influence of gravity to form a rotating disk of material that then further condensed to form the Sun and planets.[25] During the first few million years of the Solar System's history, an accretion process of sticky collisions caused the clumping of small particles, which gradually increased in size. Once the clumps reached sufficient mass, they could draw in other bodies through gravitational attraction and become planetesimals. This gravitational accretion led to the formation of the rocky planets and the gas giants.

Planetesimals within the region which would become the asteroid belt were too strongly perturbed by Jupiter's gravity to form a planet. Instead they continued to orbit the Sun as before, while occasionally colliding.[26] In regions where the average velocity of the collisions was too high, the shattering of planetesimals tended to dominate over accretion,[27] preventing the formation of planet-sized bodies. Orbital resonances occurred where the orbital period of an object in the belt formed an integer fraction of the orbital period of Jupiter, perturbing the object into a different orbit; the region lying between the orbits of Mars and Jupiter contains many such orbital resonances. As Jupiter migrated inward following its formation, these resonances would have swept across the asteroid belt, dynamically exciting the region's population and increasing their velocities relative to each other.[28]

During the early history of the Solar System, the asteroids melted to some degree, allowing elements within them to be partially or completely differentiated by mass. Some of the progenitor bodies may even have undergone periods of explosive volcanism and formed magma oceans. However, because of the relatively small size of the bodies, the period of melting was necessarily brief (compared to the much larger planets), and had generally ended about 4.5 billion years ago, in the first tens of millions of years of formation.[29] In August 2007, a study of zircon crystals in an Antarctic meteorite believed to have originated from 4 Vesta suggested that it, and by extension the rest of the asteroid belt, had formed rather quickly, within ten million years of the Solar System's origin.[30]

Evolution

The asteroids are not samples of the primordial Solar System. They have undergone considerable evolution since their formation, including internal heating (in the first few tens of millions of years), surface melting from impacts, space weathering from radiation, and bombardment by micrometeorites.[31] While some scientists refer to the asteroids as residual planetesimals,[32] other scientists consider them distinct.[33]

The current asteroid belt is believed to contain only a small fraction of the mass of the primordial belt. Computer simulations suggest that the original asteroid belt may have contained mass equivalent to the Earth. Primarily because of gravitational perturbations, most of the material was ejected from the belt within about a million years of formation, leaving behind less than 0.1% of the original mass.[26] Since their formation, the size distribution of the asteroid belt has remained relatively stable: there has been no significant increase or decrease in the typical dimensions of the main-belt asteroids.[34]

The 4:1 orbital resonance with Jupiter, at a radius 2.06 AU, can be considered the inner boundary of the asteroid belt. Perturbations by Jupiter send bodies straying there into unstable orbits. Most bodies formed inside the radius of this gap were swept up by Mars (which has an aphelion at 1.67 AU) or ejected by its gravitational perturbations in the early history of the Solar System.[35] The Hungaria asteroids lie closer to the Sun than the 4:1 resonance, but are protected from disruption by their high inclination.[36]

When the asteroid belt was first formed, the temperatures at a distance of 2.7 AU from the Sun formed a "snow line" below the freezing point of water. Planetesimals formed beyond this radius were able to accumulate ice.[37] [38] In 2006 it was announced that a population of comets had been discovered within the asteroid belt beyond the snow line, which may have provided a source of water for Earth's oceans. According to some models, there was insufficient outgassing of water during the Earth's formative period to form the oceans, requiring an external source such as a cometary bombardment.[39]

Characteristics

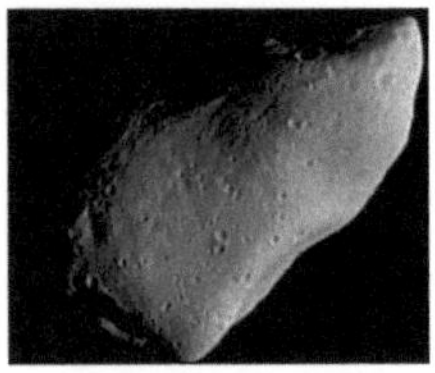

951 Gaspra, the first asteroid imaged by a spacecraft, as viewed during *Galileo*'s 1991 flyby; colors are exaggerated

Fragment of the Allende meteorite, a carbonaceous chondrite that fell to Earth in Mexico in 1969

Contrary to popular imagery, the asteroid belt is mostly empty. The asteroids are spread over such a large volume that it would be improbable to reach an asteroid without aiming carefully. Nonetheless, hundreds of thousands of asteroids are currently known, and the total number ranges in the millions or more, depending on the lower size cutoff. Over 200 asteroids are known to be larger than 100 km,[40] while a survey in the infrared wavelengths shows that the asteroid belt has 700,000 to 1.7 million asteroids with a diameter of 1 km or more.[41] The apparent magnitudes of most of the known asteroids are 11–19, with the median at about 16.[42]

The total mass of the asteroid belt is estimated to be 2.8×10^{21} to 3.2×10^{21} kilograms, which is just 4% of the mass of the Moon.[2] The four largest objects, Ceres, 4 Vesta, 2 Pallas, and 10 Hygiea, account for half of the belt's total mass, with almost one-third accounted for by Ceres alone.[3] [4]

Composition

The current belt consists primarily of three categories of asteroids: C-type or carbonaceous asteroids, S-type or silicate asteroids, and M-type or metallic asteroids.

Carbonaceous asteroids, as their name suggests, are carbon-rich and dominate the belt's outer regions.[43] Together they comprise over 75% of the visible asteroids. They are more red in hue than the other asteroids and have a very low albedo. Their surface composition is similar to carbonaceous chondrite meteorites. Chemically, their spectra match the primordial composition of the early Solar System, with only the lighter elements and volatiles removed.

S-type (silicate-rich) asteroids are more common toward the inner region of the belt, within 2.5 AU of the Sun.[43] [44] The spectra of their surfaces reveal the presence of silicates and some metal, but no significant carbonaceous compounds. This indicates that their materials have been significantly modified from their primordial composition, probably through melting and reformation. They have a relatively high albedo, and form about 17% of the total asteroid population.

M-type (metal-rich) asteroids form about 10% of the total population; their spectra resemble that of iron-nickel. Some are believed to have formed from the metallic cores of differentiated progenitor bodies that were disrupted through collision. However, there are also some silicate compounds that can produce a similar appearance. For example, the large M-type asteroid 22 Kalliope does not appear to be primarily composed of metal.[45] Within the asteroid belt, the number distribution of M-type asteroids peaks at a semi-major axis of about 2.7 AU.[46] It is not yet clear whether all M-types are compositionally similar, or whether it is a label for several varieties which do not fit neatly into the main C and S classes.[47]

One mystery of the asteroid belt is the relative rarity of V-type, or basaltic asteroids.[48] Theories of asteroid formation predict that objects the size of Vesta or larger should form crusts and mantles, which would be composed mainly of basaltic rock, resulting in more than half of all asteroids being composed either of basalt or olivine. Observations, however, suggest that 99 percent of the predicted basaltic material is missing.[49] Until 2001, most basaltic bodies discovered in the asteroid belt were believed to originate from the asteroid Vesta (hence their name V-type). However, the discovery of the asteroid 1459 Magnya revealed a slightly different chemical composition from the other basaltic asteroids discovered until then, suggesting a different origin.[49] This hypothesis was reinforced by the further discovery in 2007 of two asteroids in the outer belt, 7472 Kumakiri and (10537) 1991 RY_{16}, with differing basaltic composition that could not have originated from Vesta. These latter two are the only V-type asteroids discovered in the outer belt to date.[48]

The temperature of the asteroid belt varies with the distance from the Sun. For dust particles within the belt, typical temperatures range from 200 K (−73 °C) at 2.2 AU down to 165 K (−108 °C) at 3.2 AU[50] However, due to rotation, the surface temperature of an asteroid can vary considerably as the sides are alternately exposed to solar radiation and then to the stellar background.

Main-belt comets

Several otherwise unremarkable bodies in the outer belt show cometary activity. Since their orbits cannot be explained through capture of classical comets, it is thought that many of the outer asteroids may be icy, with the ice occasionally exposed to sublimation through small impacts. Main-belt comets may have been a major source of the Earth's oceans, since the deuterium-hydrogen ratio is too low for classical comets to have been the principal source.[51]

Orbits

Most asteroids within the asteroid belt have orbital eccentricities of less than 0.4, and an inclination of less than 30°. The orbital distribution of the asteroids reaches a maximum at an eccentricity of around 0.07 and an inclination below 4°.[42] Thus while a typical asteroid has a relatively circular orbit and lies near the plane of the ecliptic, some asteroid orbits can be highly eccentric or travel well outside the ecliptic plane.

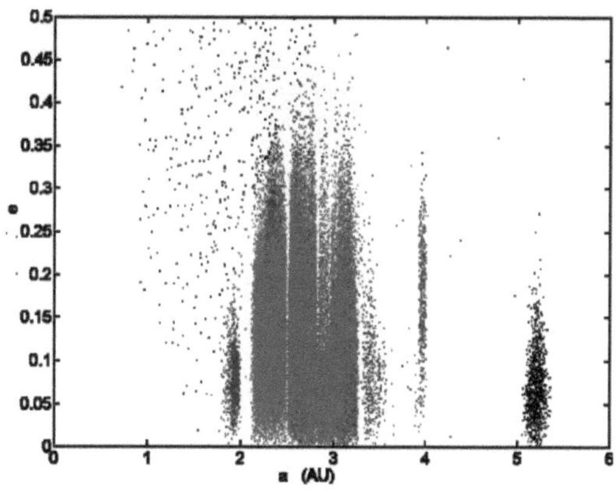

The asteroid belt (showing eccentricities), with the asteroid belt in red and blue ("core" region in red)

Sometimes, the term *main belt* is used to refer only to the more compact "core" region where the greatest concentration of bodies is found. This lies between the strong 4:1 and 2:1 Kirkwood gaps at 2.06 and 3.27 AU, and at orbital eccentricities less than roughly 0.33, along with orbital inclinations below about 20°. This "core" region contains approximately 93.4% of all numbered minor planets within the Solar System.[52]

Kirkwood gaps

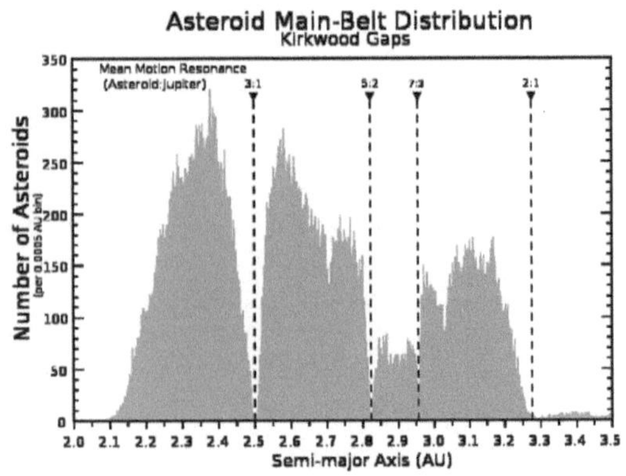

This chart shows the distribution of asteroid semi-major axes in the "core" of the asteroid belt. Black arrows point to the Kirkwood gaps, where orbital resonances with Jupiter destabilize orbits.

The semi-major axis of an asteroid is used to describe the dimensions of its orbit around the Sun, and its value determines the minor planet's orbital period. In 1866, Daniel Kirkwood announced the discovery of gaps in the distances of these bodies' orbits from the Sun. They were located at positions where their period of revolution about the Sun was an integer fraction of Jupiter's orbital period. Kirkwood proposed that the gravitational perturbations of the planet led to the removal of asteroids from these orbits.[53]

When the mean orbital period of an asteroid is an integer fraction of the orbital period of Jupiter, a mean-motion resonance with the gas giant is created that is sufficient to perturb an asteroid to new orbital elements. Asteroids that become located in the gap orbits (either primordially because of the migration of Jupiter's orbit,[54] or due to prior perturbations or collisions) are gradually nudged into different, random orbits with a larger or smaller semi-major axis.

The gaps are not seen in a simple snapshot of the locations of the asteroids at any one time because asteroid orbits are elliptical, and many asteroids still cross through the radii corresponding to the gaps. The actual spatial density of asteroids in these gaps does not differ significantly from the neighboring regions.[55]

The main gaps occur at the 3:1, 5:2, 7:3, and 2:1 mean-motion resonances with Jupiter. An asteroid in the 3:1 Kirkwood gap would orbit the Sun three times for each Jovian orbit, for instance. Weaker resonances occur at other semi-major axis values, with fewer asteroids found than nearby. (For example, an 8:3 resonance for asteroids with a semi-major axis of 2.71 AU.)[56]

The main or core population of the asteroid belt is sometimes divided into three zones, based on the most prominent Kirkwood gaps. Zone I lies between the 4:1 resonance (2.06 AU) and 3:1 resonance (2.5 AU) Kirkwood gaps. Zone II continues from the end of Zone I out to the 5:2 resonance gap (2.82 AU). Zone III extends from the outer edge of Zone II to the 2:1 resonance gap (3.28 AU).[57]

The asteroid belt may also be divided into the inner and outer belts, with the inner belt formed by asteroids orbiting nearer to Mars than the 3:1 Kirkwood gap (2.5 AU), and the outer belt formed by those asteroids closer to Jupiter's orbit. (Some authors subdivide the inner and outer belts at the 2:1 resonance gap (3.3 AU), while others suggest inner, middle, and outer belts.)

Collisions

The high population of the asteroid belt makes for a very active environment, where collisions between asteroids occur frequently (on astronomical time scales). Collisions between main-belt bodies with a mean radius of 10 km are expected to occur about once every 10 million years.[58] A collision may fragment an asteroid into numerous smaller pieces (leading to the formation of a new asteroid family). Conversely, collisions that occur at low relative speeds may also join two asteroids. After more than 4 billion years of such processes, the members of the asteroid belt now bear little resemblance to the original population.

The zodiacal light, created in part by dust from collisions in the asteroid belt

Along with the asteroid bodies, the asteroid belt also contains bands of dust with particle radii of up to a few hundred micrometres. This fine material is produced, at least in part, from collisions between asteroids, and by the impact of micrometeorites upon the asteroids. Due to the Poynting-Robertson effect, the pressure of solar radiation causes this dust to slowly spiral inward toward the Sun.[59]

The combination of this fine asteroid dust, as well as ejected cometary material, produces the zodiacal light. This faint auroral glow can be viewed at night extending from the direction of the Sun along the plane of the ecliptic. Particles that produce the visible zodiacal light average about 40 μm in radius. The typical lifetimes of such particles are about 700,000 years. Thus, to maintain the bands of dust, new particles must be steadily produced within the asteroid belt.[59]

Meteorites

Some of the debris from collisions can form meteoroids that enter the Earth's atmosphere.[60] More than 99.8 percent of the 30,000 meteorites found on Earth to date are believed to have originated in the asteroid belt.[61] A September 2007 study by a joint US-Czech team has suggested that a large-body collision undergone by the asteroid 298 Baptistina sent a number of fragments into the inner solar system. The impacts of these fragments are believed to have created both Tycho crater on the Moon and Chicxulub crater in Mexico, the relict of the massive impact which is believed to have triggered the extinction of the dinosaurs 65 million years ago.[62]

Families and groups

In 1918, the Japanese astronomer Kiyotsugu Hirayama noticed that the orbits of some of the asteroids had similar parameters, forming families or groups.[63]

Approximately one-third of the asteroids in the asteroid belt are members of an asteroid family. These share similar orbital elements, such as semi-major axis, eccentricity, and orbital inclination as well as similar spectral features, all of which indicate a common origin in the breakup of a larger body. Graphical displays of these elements, for members of the asteroid belt, show concentrations indicating the presence of an asteroid family. There are about 20–30 associations that are almost certainly asteroid families. Additional groupings have been found that are less certain. Asteroid families can be confirmed when the members display common spectral features.[64] Smaller associations of asteroids are called groups or clusters.

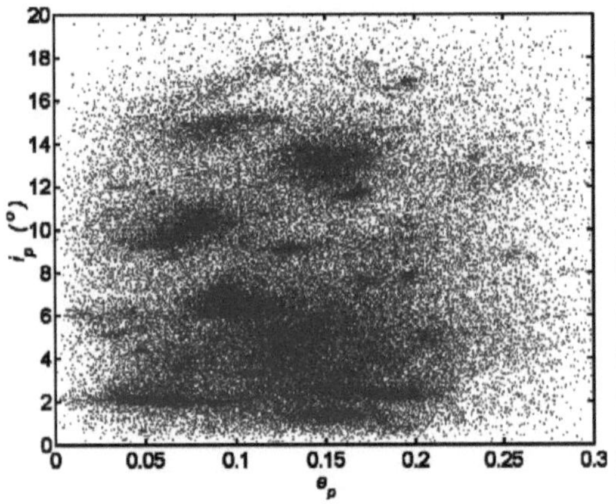

This plot of orbital inclination (i_p) versus eccentricity (e_p) for the numbered main-belt asteroids clearly shows clumpings representing asteroid families.

Some of the most prominent families in the asteroid belt (in order of increasing semi-major axes) are the Flora, Eunoma, Koronis, Eos, and Themis families.[46] The Flora family, one of the largest with more than 800 known members, may have formed from a collision less than a billion years ago.[65] The largest asteroid to be a true member of a family (as opposed to an interloper in the case of Ceres with the Gefion family) is 4 Vesta. The Vesta family is believed to have formed as the result of a crater-forming impact on Vesta. Likewise, the HED meteorites may also have originated from Vesta as a result of this collision.[66]

Three prominent bands of dust have been found within the asteroid belt. These have similar orbital inclinations as the Eos, Koronis, and Themis asteroid families, and so are possibly associated with those groupings.[67]

Periphery

Skirting the inner edge of the belt (ranging between 1.78 and 2.0 AU, with a mean semi-major axis of 1.9 AU) is the Hungaria family of minor planets. They are named after the main member, 434 Hungaria; the group contains at least 52 named asteroids. The Hungaria group is separated from the main body by the 4:1 Kirkwood gap and their orbits have a high inclination. Some members belong to the Mars-crossing category of asteroids, and gravitational perturbations by Mars are likely a factor in reducing the total population of this group.[68]

Another high-inclination group in the inner part of the asteroid belt is the Phocaea family. These are composed primarily of S-type asteroids, whereas the neighboring Hungaria family includes some E-types.[69] The Phocaea family orbit between 2.25 and 2.5 AU from the Sun.

Skirting the outer edge of the asteroid belt is the Cybele group, orbiting between 3.3 and 3.5 AU. These have a 7:4 orbital resonance with Jupiter. The Hilda family orbit between 3.5 and 4.2 AU, and have relatively circular orbits and a stable 3:2 orbital resonance with Jupiter. There are few asteroids beyond 4.2 AU, until Jupiter's orbit. Here the two families of Trojan asteroids can be found, which, at least for objects larger than 1 km, are approximately as numerous as the asteroids of the asteroid belt.[70]

New families

Some asteroid families have formed recently, in astronomical terms. The Karin Cluster apparently formed about 5.7 million years ago from a collision with a 33 km radius progenitor asteroid.[71] The Veritas family formed about 8.3 million years ago; evidence includes interplanetary dust recovered from ocean sediment.[72]

More recently, the Datura cluster appears to have formed about 450 thousand years ago from a collision with a main-belt asteroid. The age estimate is based on the probability of the members having their current orbits, rather than from any physical evidence. However, this cluster may have been a source for some zodiacal dust material.[73] Other recent cluster formations, such as the Iannini cluster (*circa* 1–5 million years ago), may have provided additional sources of this asteroid dust.[74]

Exploration

The first spacecraft to traverse the asteroid belt was Pioneer 10, which entered the region on July 16, 1972. At the time there was some concern that the debris in the belt would pose a hazard to the spacecraft, but it has since been safely traversed by 9 Earth-based craft without incident. Pioneer 11, Voyagers 1 and 2 and Ulysses passed through the belt without imaging any asteroids. Galileo imaged the asteroid 951 Gaspra in 1991 and 243 Ida in 1993, NEAR imaged 253 Mathilde in 1997, Cassini imaged 2685 Masursky in 2000, Stardust imaged 5535 Annefrank in 2002, New Horizons imaged 132524 APL in 2006, and Rosetta imaged 2867 Šteins in 2008.[75] Due to the low density of materials within the belt, the odds of a probe running into an asteroid are now estimated at less than one in a billion.[76]

Artist's concept of the Dawn Mission spacecraft with Vesta (left) and Ceres (right)

All spacecraft images of belt asteroids to date have come from brief flyby opportunities by probes headed for other targets. Only the NEAR and Hayabusa missions have studied asteroids for a protracted period in orbit and at the surface and these were near-Earth asteroids. However, the Dawn Mission has been dispatched to explore Vesta and Ceres in the asteroid belt. If the probe is still operational after examining these two large bodies, an extended mission is possible that could allow additional exploration, possibly of Pallas.[77]

See also

- Asteroid mining
- Asteroids in astrology
- Asteroids in fiction
- Colonization of the asteroids
- Debris disk
- Kuiper belt

References

[1] Krasinsky, G. A.; Pitjeva, E. V.; Vasilyev, M. V.; Yagudina, and E. I. (July 2002). "Hidden Mass in the Asteroid Belt". *Icarus* **158** (1): 98–105. Bibcode 2002Icar..158...98K. doi:10.1006/icar.2002.6837.

[2] Pitjeva, E. V. (2005). "High-Precision Ephemerides of Planets—EPM and Determination of Some Astronomical Constants" (http://iau-comm4.jpl.nasa.gov/EPM2004.pdf) (PDF). *Solar System Research* **39** (3): 176. Bibcode 2005SoSyR..39..176P. doi:10.1007/s11208-005-0033-2. .

[3] For recent estimates of the masses of Ceres, 4 Vesta, 2 Pallas and 10 Hygiea, see the references in the infoboxes of their respective articles.

[4] Yeomans, Donald K. (July 13, 2006). "JPL Small-Body Database Browser" (http://ssd.jpl.nasa.gov/sbdb.cgi). NASA JPL. . Retrieved 2010-09-27.

[5] Hilton, J. (2001). "When Did the Asteroids Become Minor Planets?" (http://www.usno.navy.mil/USNO/astronomical-applications/astronomical-information-center/minor-planets). *US Naval Observatory (USNO)*. . Retrieved 2007-10-01.

[6] "Dawn: A Journey to the Beginning of the Solar System" (http://www-ssc.igpp.ucla.edu/dawn/background.html). *Space Physics Center: UCLA*. 2005. . Retrieved 2007-11-03.

[7] Hoskin, Michael. "Bode's Law and the Discovery of Ceres" (http://www.astropa.unipa.it/HISTORY/hoskin.html). *Churchill College, Cambridge*. . Retrieved 2010-07-12.

[8] Linda T. Elkins-Tanton, *Asteroids, Meteorites, and Comets*, 2010:10

[9] "Call the police! The story behind the discovery of the asteroids". *Astronomy Now* (June 2007): 60–61.

[10] Pogge, Richard (2006). "An Introduction to Solar System Astronomy: Lecture 45: Is Pluto a Planet?" (http://www.astronomy.ohio-state.edu/~pogge/Ast161/Unit6/dwarfs.html). *An Introduction to Solar System Astronomy*. Ohio State University. . Retrieved 2007-11-11.

[11] Harper, Douglas (2010). "Asteroid" (http://www.etymonline.com/index.php?search=asteroid&searchmode=none). *Online Etymology Dictionary*. Etymology Online. . Retrieved 2011-04-15.

[12] DeForest, Jessica (2000). "Greek and Latin Roots" (http://www.msu.edu/~defores1/gre/roots/gre_rts_afx2.htm). Michigan State University. . Retrieved 2007-07-25.

[13] Cunningham, Clifford (1984). "William Hershel and the First Two Asteroids". *The Minor Planet Bulletin* **11**: 3. Bibcode 1984MPBu...11....3C.

[14] Staff (2002). "Astronomical Serendipity" (http://dawn.jpl.nasa.gov/DawnCommunity/flashbacks/fb_06.asp). NASA JPL. . Retrieved 2007-04-20.

[15] "Is it a coincidence that most of the planets fall within the Titius-Bode law's boundaries?" (http://www.astronomy.com/asy/default.aspx?c=a&id=4494). *astronomy.com*. . Retrieved 2007-10-16.

[16] von Humboldt, Alexander (1850). *Cosmos: A Sketch of a Physical Description of the Universe*. **1**. Harper & Brothers, New York (NY). p. 44. ISBN 0-8018-5503-9.

[17] Mann, Robert James (1852). *A Guide to the Knowledge of the Heavens*. Jarrold. p. 171. and 1853, p. 216

[18] "Further Investigation relative to the form, the magnitude, the mass, and the orbit of the Asteroid Planets" (http://books.google.com/?id=hhQAAAAAMAAJ&pg=PA191&dq=asteroid+belt). *The Edinburgh New Philosophical Journal* **5**: 191. January–April 1857. .: "[Professor Peirce] then observed that the analogy between the ring of Saturn and the belt of the asteroids was worthy of notice."

[19] Hughes, David W. (2007). "A Brief History of Asteroid Spotting" (http://www.open2.net/sciencetechnologynature/planetsbeyond/asteroids/history.html). BBC. . Retrieved 2007-04-20.

[20] Moore, Patrick; Rees, Robin (2011). *Patrick Moore's Data Book of Astronomy* (2nd ed.). Cambridge University Press. p. 156. ISBN 0521899354.

[21] Manley, Scott (August 25, 2010). "Asteroid Discovery from 1980 to 2010" (http://www.youtube.com/watch?v=S_d-gs0WoUw). *You Tube*. . Retrieved 2011-04-15.

[22] "MPC Archive Statistics" (http://www.minorplanetcenter.org/iau/lists/ArchiveStatistics.html). IAU Minor Planet Center. . Retrieved 2011-04-04.

[23] "A Brief History of Asteroid Spotting" (http://www.open2.net/sciencetechnologynature/planetsbeyond/asteroids/history.html). *Open2.net*. . Retrieved 2007-05-15.

[24] Masetti, M.; and Mukai, K. (December 1, 2005). "Origin of the Asteroid Belt" (http://imagine.gsfc.nasa.gov/docs/ask_astro/answers/980810a.html). NASA Goddard Spaceflight Center. . Retrieved 2007-04-25.

[25] Watanabe, Susan (July 20, 2001). "Mysteries of the Solar Nebula" (http://www.jpl.nasa.gov/news/features.cfm?feature=520). NASA. . Retrieved 2007-04-02.

[26] Petit, J.-M.; Morbidelli, A.; and Chambers, J. (2001). "The Primordial Excitation and Clearing of the Asteroid Belt" (http://www.gps.caltech.edu/classes/ge133/reading/asteroids.pdf) (PDF). *Icarus* **153** (2): 338–347. Bibcode 2001Icar..153..338P. doi:10.1006/icar.2001.6702. . Retrieved 2007-03-22.

[27] Edgar, R.; and Artymowicz, P. (2004). "Pumping of a Planetesimal Disc by a Rapidly Migrating Planet" (http://www.astro.su.se/~pawel/edgar+artymowicz.pdf) (PDF). *Monthly Notices of the Royal Astronomical Society* **354** (3): 769–772. arXiv:astro-ph/0409017. Bibcode 2004MNRAS.354..769E. doi:10.1111/j.1365-2966.2004.08238.x. . Retrieved 2007-04-16.

[28] Scott, E. R. D. (March 13–17, 2006). "Constraints on Jupiter's Age and Formation Mechanism and the Nebula Lifetime from Chondrites and Asteroids" (http://adsabs.harvard.edu/abs/2006LPI....37.2367S). *Proceedings 37th Annual Lunar and Planetary Science Conference*. League City, Texas: Lunar and Planetary Society. . Retrieved 2007-04-16.

[29] Taylor, G. J.; Keil, K.; McCoy, T.; Haack, H.; and Scott, E. R. D.; Keil; McCoy; Haack; Scott (1993). "Asteroid differentiation – Pyroclastic volcanism to magma oceans". *Meteoritics* **28** (1): 34–52. Bibcode 1993Metic..28...34T.

[30] Kelly, Karen (2007). "U of T researchers discover clues to early solar system" (http://webapps.utsc.utoronto.ca/ose/story.php?id=665). *University of Toronto*. . Retrieved 2010-07-12.

[31] Clark, B. E.; Hapke, B.; Pieters, C.; Britt, D.; Hapke; Pieters; Britt (2002). "Asteroid Space Weathering and Regolith Evolution". *Asteroids III*: 585. Bibcode 2002aste.conf..585C. Gaffey, Michael J. (1996). "The Spectral and Physical Properties of Metal in Meteorite Assemblages: Implications for Asteroid Surface Materials". *Icarus (ISSN 0019-1035)* **66** (3): 468. Bibcode 1986Icar...66..468G. doi:10.1016/0019-1035(86)90086-2. Keil, K. (2000). "Thermal alteration of asteroids: evidence from meteorites" (http://www.ingentaconnect.com/content/els/00320633/2000/00000048/00000010/art00054). *Planetary and Space Science*. . Retrieved 2007-11-08. Baragiola, R. A.; Duke, C. A.; Loeffler, M.; McFadden, L. A.; and Sheffield, J.; Duke; Loeffler; McFadden; Sheffield (2003). "Impact of ions and micrometeorites on mineral surfaces: Reflectance changes and production of atmospheric species in airless solar system bodies". *EGS - AGU - EUG Joint Assembly*: 7709. Bibcode 2003EAEJA.....7709B.

[32] Chapman, C. R.; Williams, J. G.; Hartmann, W. K. (1978). "The asteroids". *Annual review of astronomy and astrophysics* **16**: 33–75. Bibcode 1978ARA&A..16...33C. doi:10.1146/annurev.aa.16.090178.000341.

[33] Kracher, A. (2005). "Asteroid 433 Eros and partially differentiated planetesimals: bulk depletion versus surface depletion of sulfur" (http://www.cosis.net/abstracts/EGU05/03788/EGU05-J-03788.pdf) (PDF). *Ames Laboratory*. . Retrieved 2007-11-08.

[34] Stiles, Lori (September 15, 2005). "Asteroids Caused the Early Inner Solar System Cataclysm" (http://uanews.org/cgi-bin/WebObjects/UANews.woa/7/wa/SRStoryDetails?ArticleID=11692). University of Arizona News. . Retrieved 2007-04-18.

[35] Alfvén, H.; Arrhenius, G. (1976). "The Small Bodies" (http://history.nasa.gov/SP-345/ch4.htm). *SP-345 Evolution of the Solar System*. NASA. . Retrieved 2007-04-12.

[36] Spratt, Christopher E. (April 1990). "The Hungaria group of minor planets". *Journal of the Royal Astronomical Society of Canada* **84**: 123–131. Bibcode 1990JRASC..84..123S.

[37] Lecar, M.; Podolak, M.; Sasselov, D.; Chiang, E. (2006). "Infrared cirrus – New components of the extended infrared emission". *The Astrophysical Journal* **640** (2): 1115–1118. arXiv:astro-ph/0602217. Bibcode 2006ApJ...640.1115L. doi:10.1086/500287.

[38] Berardelli, Phil (March 23, 2006). "Main-Belt Comets May Have Been Source Of Earths Water" (http://www.spacedaily.com/reports/Main_Belt_Comets_May_Have_Been_Source_Of_Earths_Water.html). Space Daily. . Retrieved 2007-10-27.

[39] Lakdawalla, Emily (April 28, 2006). "Discovery of a Whole New Type of Comet" (http://www.planetary.org/blog/article/00000551/). The Planetary Society. . Retrieved 2007-04-20.

[40] Yeomans, Donald K. (April 26, 2007). "JPL Small-Body Database Search Engine" (http://ssd.jpl.nasa.gov/sbdb_query.cgi). NASA JPL. . Retrieved 2007-04-26. – search for asteroids in the main belt regions with a diameter >100.

[41] Tedesco, E. F.; and Desert, F.-X. (2002). "The Infrared Space Observatory Deep Asteroid Search". *The Astronomical Journal* **123** (4): 2070–2082. Bibcode 2002AJ....123.2070T. doi:10.1086/339482.

[42] Williams, Gareth (September 25, 2010). "Distribution of the Minor Planets" (http://www.minorplanetcenter.org/iau/lists/MPDistribution.html). Minor Planets Center. . Retrieved 2010-10-27.

[43] Wiegert, P.; Balam, D.; Moss, A.; Veillet, C.; Connors, M.; and Shelton, I. (2007). "Evidence for a Color Dependence in the Size Distribution of Main-Belt Asteroids" (http://astro.uwo.ca/~wiegert/papers/2007AJ.133.1609.pdf). *The Astronomical Journal* **133** (4): 1609–1614. arXiv:astro-ph/0611310. Bibcode 2007AJ....133.1609W. doi:10.1086/512128. . Retrieved 2008-09-06.

[44] Clark, B. E. (1996). "New News and the Competing Views of Asteroid Belt Geology". *Lunar and Planetary Science* **27**: 225–226. Bibcode 1996LPI....27..225C.

[45] Margot, J. L.; and Brown, M. E. (2003). "A Low-Density M-type Asteroid in the Main Belt". *Science* **300** (5627): 1939–1942. Bibcode 2003Sci...300.1939M. doi:10.1126/science.1085844. PMID 12817147.

[46] Lang, Kenneth R. (2003). "Asteroids and meteorites" (http://ase.tufts.edu/cosmos/print_images.asp?id=15). NASA's Cosmos. . Retrieved 2007-04-02.

[47] Mueller, M.; Harris, A. W.; Delbo, M.; and the MIRSI Team; Harris; Delbo (2005). "21 Lutetia and other M-types: Their sizes, albedos, and thermal properties from new IRTF measurements". *Bulletin of the American Astronomical Society* **37**: 627. Bibcode 2005DPS....37.0702M.

[48] Duffard, R. D.; Roig, F. (July 14–18, 2008). "Two New Basaltic Asteroids in the Main Belt?". *Asteroids, Comets, Meteors 2008*. Baltimore, Maryland. arXiv:0704.0230. Bibcode 2008LPICo1405.8154D.

[49] Than, Ker (2007). "Strange Asteroids Baffle Scientists" (http://www.space.com/scienceastronomy/070821_basalt_asteroid.html). *space.com*. . Retrieved 2007-10-14.

[50] Low, F. J.; *et al.* (1984). "Infrared cirrus – New components of the extended infrared emission". *Astrophysical Journal, Part 2 – Letters to the Editor* **278**: L19–L22. Bibcode 1984ApJ...278L..19L. doi:10.1086/184213.

[51] "Interview with David Jewitt" (http://www.youtube.com/watch?v=B1W4NTmI5Bk). Youtube.com. 2007-01-05. . Retrieved 2011-05-21.

[52] This value was obtained by a simple count up of all bodies in that region using data for 120437 numbered minor planets from the Minor Planet Center orbit database (http://www.minorplanetcenter.org/iau/MPCORB.html), dated February 8, 2006.

[53] Fernie, J. Donald (1999). "The American Kepler" (http://www.americanscientist.org/issues/pub/1999/9/the-american-kepler/2). *The Americal Scientist* **87** (5): 398. . Retrieved 2007-02-04.

[54] Liou, Jer-Chyi; and Malhotra, Renu (1997). "Depletion of the Outer Asteroid Belt" (http://www.sciencemag.org/cgi/content/full/275/5298/375). *Science* **275** (5298): 375–377. Bibcode 1997Sci...275..375L. doi:10.1126/science.275.5298.375. PMID 8994031. . Retrieved 2007-08-01.

[55] McBride, N.; and Hughes, D. W.; Hughes (1990). "The spatial density of asteroids and its variation with asteroidal mass". *Monthly Notices of the Royal Astronomical Society* **244**: 513–520. Bibcode 1990MNRAS.244..513M.

[56] Ferraz-Mello, S. (June 14–18, 1993). "Kirkwood Gaps and Resonant Groups" (http://adsabs.harvard.edu/abs/1994IAUS..160..175F). *proceedings of the 160th International Astronomical Union.* Belgirate, Italy: Kluwer Academic Publishers. pp. 175–188. . Retrieved 2007-03-28.

[57] Klacka, Jozef (1992). "Mass distribution in the asteroid belt". *Earth, Moon, and Planets* **56** (1): 47–52. Bibcode 1992EM&P...56...47K. doi:10.1007/BF00054599.

[58] Backman, D. E. (March 6, 1998). "Fluctuations in the General Zodiacal Cloud Density" (http://astrobiology.arc.nasa.gov/workshops/zodiac/backman/backman_toc.html). *Backman Report.* NASA Ames Research Center. . Retrieved 2007-04-04.

[59] Reach, William T. (1992). "Zodiacal emission. III – Dust near the asteroid belt". *Astrophysical Journal* **392** (1): 289–299. Bibcode 1992ApJ...392..289R. doi:10.1086/171428.

[60] Kingsley, Danny (May 1, 2003). "Mysterious meteorite dust mismatch solved" (http://abc.net.au/science/news/stories/s843594.htm). ABC Science. . Retrieved 2007-04-04.

[61] "Meteors and Meteorites" (http://www.nasa.gov/pdf/145945main_ Meteors.Meteorites.Lithograph.pdf). NASA. . Retrieved 2010-07-12.

[62] "Breakup event in the main asteroid belt likely caused dinosaur extinction 65 million years ago" (http://www.physorg.com/news108218928.html). *Southwest Research Institute.* 2007. . Retrieved 2007-10-14.

[63] Hughes, David W. (2007). "Finding Asteroids In Space" (http://www.open2.net/sciencetechnologynature/planetsbeyond/asteroids/finding.html). BBC. . Retrieved 2007-04-20.

[64] Lemaitre, Anne (August 31-September 4, 2004). "Asteroid family classification from very large catalogues" (http://adsabs.harvard.edu/abs/2005dpps.conf..135L). *Proceedings Dynamics of Populations of Planetary Systems.* Belgrade, Serbia and Montenegro: Cambridge University Press. pp. 135–144. . Retrieved 2007-04-15.

[65] Martel, Linda M. V. (March 9, 2004). "Tiny Traces of a Big Asteroid Breakup" (http://www.psrd.hawaii.edu/Mar04/fossilMeteorites.html). Planetary Science Research Discoveries. . Retrieved 2007-04-02.

[66] Drake, Michael J. (2001). "The eucrite/Vesta story". *Meteoritics & Planetary Science* **36** (4): 501–513. Bibcode 2001M&PS...36..501D. doi:10.1111/j.1945-5100.2001.tb01892.x.

[67] Love, S. G.; and Brownlee, D. E. (1992). "The IRAS dust band contribution to the interplanetary dust complex – Evidence seen at 60 and 100 microns". *Astronomical Journal* **104** (6): 2236–2242. Bibcode 1992AJ....104.2236L. doi:10.1086/116399.

[68] Spratt, Christopher E. (1990). "The Hungaria group of minor planets". *Journal of the Royal Astronomical Society of Canada* **84** (2): 123–131. Bibcode 1990JRASC..84..123S.

[69] Carvano, J. M.; Lazzaro, D.; Mothé-Diniz, T.; Angeli, C. A.; and Florczak, M. (2001). "Spectroscopic Survey of the Hungaria and Phocaea Dynamical Groups". *Icarus* **149** (1): 173–189. Bibcode 2001Icar..149..173C. doi:10.1006/icar.2000.6512.

[70] Dymock, Roger (2010). *Asteroids and Dwarf Planets and How to Observe Them* (http://books.google.com/books?id=vQcAnwt_87sC&pg=PA24). Astronomers' Observing Guides. Springer. p. 24. ISBN 144196438X. . Retrieved 2011-04-04.

[71] Nesvorný, David; *et al.* (August 2006). "Karin cluster formation by asteroid impact". *Icarus* **183** (2): 296–311. Bibcode 2006Icar..183..296N. doi:10.1016/j.icarus.2006.03.008.

[72] McKee, Maggie (January 18, 2006). "Eon of dust storms traced to asteroid smash" (http://space.newscientist.com/channel/solar-system/comets-asteroids/dn8603). New Scientist Space. . Retrieved 2007-04-15.

[73] Nesvorný, D.; Vokrouhlick, D.; and Bottke, W. F. (2006). "The Breakup of a Main-Belt Asteroid 450 Thousand Years Ago" (http://www.sciencemag.org/cgi/content/full/312/5779/1490). *Science* **312** (5779): 1490. Bibcode 2006Sci...312.1490N. doi:10.1126/science.1126175. PMID 16763141. . Retrieved 2007-04-15.

[74] Nesvorný, D.; Bottke, W. F.; Levison, H. F.; and Dones, L. (2003). "Recent Origin of the Solar System Dust Bands" (http://iopscience.iop.org/0004-637X/591/1/486/pdf/0004-637X_591_1_486.pdf). *The Astrophysical Journal* **591** (1): 486–497. Bibcode 2003ApJ...591..486N. doi:10.1086/374807. . Retrieved 2007-04-15.

[75] Barucci, M. A.; Fulchignoni, M.; and Rossi, A. (2007). "Rosetta Asteroid Targets: 2867 Steins and 21 Lutetia". *Space Science Reviews* **128** (1–4): 67–78. Bibcode 2007SSRv..128...67B. doi:10.1007/s11214-006-9029-6.

[76] Stern, Alan (June 2, 2006). "New Horizons Crosses The Asteroid Belt" (http://www.spacedaily.com/reports/New_Horizons_Crosses_The_Asteroid_Belt.html). Space Daily. . Retrieved 2007-04-14.

[77] Staff (April 10, 2007). "Dawn Mission Home Page" (http://dawn.jpl.nasa.gov/). NASA JPL. . Retrieved 2007-04-14.

Further reading

- Elkins-Tanton, Linda T. (2006). *Asteroids, Meteorites, and Comets* (First ed.). New York: Chelsea House. ISBN 0-8160-5195-X.

External links

- Asteroid Discovery from 1980 to 2010 (http://www.youtube.com/watch?v=S_d-gs0WoUw)
- Arnett, William A. (February 26, 2006). "Asteroids" (http://www.nineplanets.org/asteroids.html). The Nine Planets. Retrieved 2007-04-20.
- Asteroids Page (http://solarsystem.nasa.gov/planets/profile.cfm?Object=Asteroids) at NASA's Solar System Exploration (http://solarsystem.nasa.gov)
- Cain, Fraser. "The Asteroid Belt" (http://www.astronomycast.com/astronomy/episode-55-the-asteroid-belt/). Universe Today. Retrieved 2008-04-01.
- Hsieh, Henry H. (March 1, 2006). "Main-Belt Comets" (http://star.pst.qub.ac.uk/~hhh/mbcs.shtml). University of Hawaii. Retrieved 2007-04-20.
- "Main Asteroid Belt" (http://www.solstation.com/stars/asteroid.htm). Sol Company. Retrieved 2007-04-20.
- Munsell, Kirk (September 16, 2005). "Asteroids: Overview" (http://solarsystem.nasa.gov/planets/profile.cfm?Object=Asteroids). NASA's Solar System Exploration (http://solarsystem.nasa.gov). Retrieved 2007-05-26.
- Plots of eccentricity vs. semi-major axis (http://burro.astr.cwru.edu/stu/media/asteroid_all_axisvecc.jpg) and inclination vs. semi-major axis (http://burro.astr.cwru.edu/stu/media/asteroid_all_axisvincl.jpg) at Asteroid Dynamic Site
- Staff (October 31, 2006). "Asteroids" (http://nssdc.gsfc.nasa.gov/planetary/planets/asteroidpage.html). NASA. Retrieved 2007-04-20.
- Staff (2007). "Space Topics: Asteroids and Comets" (http://www.planetary.org/explore/topics/asteroids_and_comets/facts.html). The Planetary Society. Retrieved 2007-04-20.

Asteroid_family

An **asteroid family** is a population of asteroids that share similar proper orbital elements, such as semimajor axis, eccentricity, and orbital inclination. The members of the families are thought to be fragments of past asteroid collisions.

General properties

Large prominent families contain several hundred recognized asteroids (and many more smaller objects which may be either not-yet-analyzed, or not-yet-discovered). Small, compact families can have only about ten identified members. About 33% to 35% of asteroids in the main belt are family members.

There are about 20 to 30 reliably recognized families, with several tens of less certain groupings. Most asteroid families are found in the main asteroid belt, although several family-like groups such as the Pallas family, Hungaria family, and the Phocaea family lie at smaller semi-major axis or larger inclination than the main belt.

One family has been identified associated with the dwarf planet Haumea.[1] Some studies have tried to find evidence of collisional families among the trojan asteroids, but at present the evidence is inconclusive.

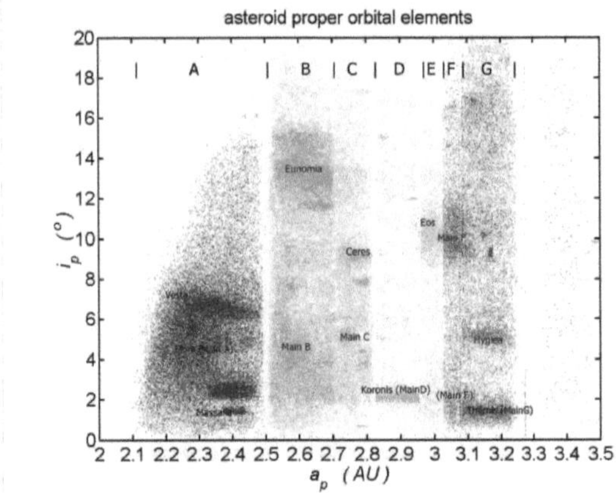

Plot of proper inclination vs. semi-major axis for numbered asteroids. Asteroid families are visible as distinct clumps. Prominent Kirkwood gaps divide the core region. (A, B+C, D, E+F+G)

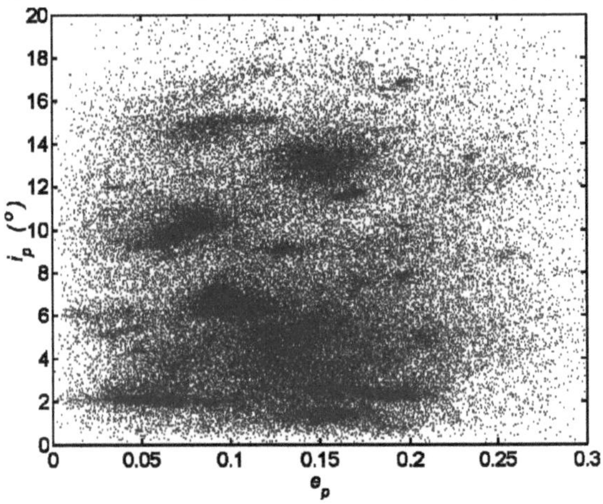

Plot of proper inclination vs. eccentricity for numbered asteroids.

Origin and evolution

The families are thought to form as a result of collisions between asteroids. In many or most cases the parent body was shattered, but there are also several families which resulted from a large cratering event which did not disrupt the parent body (e.g. the Vesta, Pallas, Hygiea, and Massalia families). Such *cratering families* typically consist of a single large body and a swarm of asteroids that are much smaller. Some families (e.g. the Flora family) have complex internal structures which are not satisfactorily explained at the moment, but may be due to several collisions in the same region at different times.

Due to the method of origin, all the members have closely matching compositions for most families. Notable exceptions are those families (such as the Vesta family) which formed from a large differentiated parent body.

Asteroid families are thought to have lifetimes of the order of a billion years, depending on various factors (e.g. smaller asteroids are lost faster). This is significantly shorter than the Solar System's age, so few if any are relics of the early Solar System. Decay of families occurs both because of slow dissipation of the orbits due to perturbations from Jupiter or other large bodies, and because of collisions between asteroids which grind them down to small bodies. Such small asteroids then become subject to perturbations such as the Yarkovsky effect that can push them towards orbital resonances with Jupiter over time. Once there, they are relatively rapidly ejected from the asteroid belt. Tentative age estimates have been obtained for some families, ranging from hundreds of millions of years to less than several million years for e.g. the compact Karin family. Old families are thought to contain few small members, and this is the basis of the age determinations.

It is supposed that many very old families have lost all the smaller and medium-sized members, leaving only a few of the largest intact. A suggested example of such old family remains are the 9 Metis and 113 Amalthea pair. Further evidence for a large number of past families (now dispersed) comes from analysis of chemical ratios in iron meteorites. These show that there must have once been at least 50 to 100 parent bodies large enough to be differentiated, that have since been shattered to expose their cores and produce the actual meteorites (Kelley & Gaffey 2000).

Identification of members and interlopers

When the orbital elements of main belt asteroids are plotted (typically inclination vs. eccentricity, or vs. semi-major axis), a number of distinct concentrations are seen against the rather uniform background distribution of generic asteroids. These concentrations are the asteroid families.

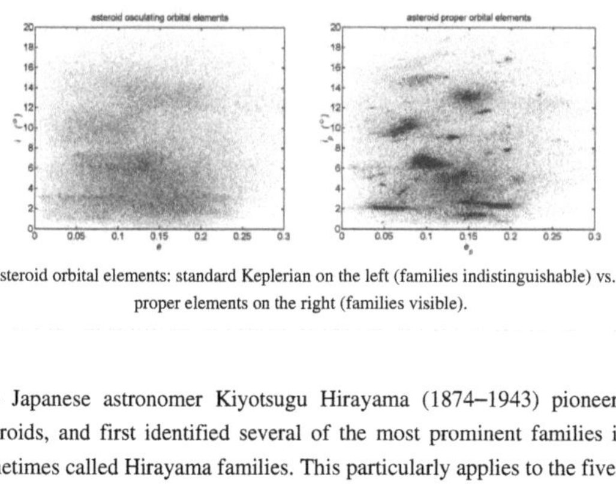

Asteroid orbital elements: standard Keplerian on the left (families indistinguishable) vs. proper elements on the right (families visible).

Strictly speaking, families and their membership are identified by analysing the so-called proper orbital elements rather than the current osculating orbital elements, which regularly fluctuate on timescales of tens of thousands of years. The *proper elements* are related constants of motion that remain almost constant for times of at least tens of millions of years, and perhaps longer.

The Japanese astronomer Kiyotsugu Hirayama (1874–1943) pioneered the estimation of proper elements for asteroids, and first identified several of the most prominent families in 1918. In his honor, asteroid families are sometimes called Hirayama families. This particularly applies to the five prominent groupings discovered by him.

Present day computer-assisted searches have identified several tens of asteroid families. The most prominent algorithms have been the Hierarchical Clustering Method (HCM) which looks for groupings with small nearest-neighbour distances in orbital element space, and the Wavelet Analysis Method (WAM) which builds a density-of-asteroids map in orbital element space, and looks for density peaks.

The boundaries of the families are somewhat vague because at the edges they blend into the background density of asteroids in the main belt. For this reason the number of members even among discovered asteroids is usually only known approximately, and membership is uncertain for asteroids near the edges.

Additionally, some *interlopers* from the heterogeneous background asteroid population are expected even in the central regions of a family. Since the true family members caused by the collision are expected to have similar compositions, most such interlopers can in principle be recognised by spectral properties which do not match those of the bulk of family members. A prominent example is 1 Ceres, the largest asteroid, which is an interloper in the family once named after it (the Ceres family, now the Gefion family).

Spectral characteristics can also be used to determine the membership (or otherwise) of asteroids in the outer regions of a family, as has been used e.g. for the Vesta family, whose members have an unusual composition.

Family types

As previously mentioned, families caused by an impact that did not disrupt the parent body but only ejected fragments are called *cratering families*. Other terminology has been used to distinguish various types of groups which are less distinct or less statistically certain from the most prominent "nominal families" (or *clusters*). The term *cluster* is also used to describe a small asteroid family, such as the Karin Cluster.[2] *Clumps* are groupings which have relatively few members but are clearly distinct from the background (e.g. the Juno clump). *Clans* are groupings which merge very gradually into the background density and/or have a complex internal structure making it difficult to decide whether they are one complex group or several unrelated overlapping groups (e.g. the Flora family has been called a clan). *Tribes* are groups that are less certain to be statistically significant against the background either because of small density or large uncertainty in the orbital parameters of the members.

List of families

Family Name	Named After	orbital elements			Size		Alternate Names
		a (AU)	e	*i* (°)	approx. % of asteroids	members in Zappalà HCM analysis[A]	
The most prominent families within the main belt are:							
Eos	221 Eos	2.99 to 3.03	0.01 to 0.13	8 to 12		480	
Eunomia	15 Eunomia	2.53 to 2.72	0.08 to 0.22	11.1 to 15.8	5%	370	
Flora	8 Flora	2.15 to 2.35	0.03 to 0.23	1.5 to 8.0	4-5%	590	Ariadne family after 43 Ariadne
Hygiea	10 Hygiea	3.06 to 3.24	0.09 to 0.19	3.5 to 6.8	1%	105	
Koronis	158 Koronis	2.83 to 2.91	0 to 0.11	0 to 3.5		310	
Maria	170 Maria	2.5 to 2.706		12 to 17		80	
Nysa	44 Nysa	2.41 to 2.5	0.12 to 0.21	1.5 to 4.3		380	Hertha family after 135 Hertha
Themis	24 Themis	3.08 to 3.24	0.09 to 0.22	0 to 3		530	
Vesta	4 Vesta	2.26 to 2.48	0.03 to 0.16	5.0 to 8.3	6%	240	
Other notable main belt families:[C]							
Adeona	145 Adeona					65	
Astrid	1128 Astrid					11	

Bower	1639 Bower					13	Endymion family after 342 Endymion
Brasilia	293 Brasilia					14	
Gefion	1272 Gefion	2.74 to 2.82	0.08 to 0.18	7.4 to 10.5	0.8%	89	Ceres family after 1 Ceres and Minerva family after 93 Minerva
Chloris	410 Chloris					24	
Dora	668 Dora					78	
Erigone	163 Erigone					47	
Hansa[3]	480 Hansa	~2.66	~0.06	~22.0°			
Hilda	153 Hilda	3.7 to 4.2	>0.07	<20°	-		
Karin	832 Karin					39[B]	
Lydia	110 Lydia					38	
Massalia	20 Massalia	2.37 to 2.45	0.12 to 0.21	0.4 to 2.4	0.8%	47	
Meliboea	137 Meliboea					15	
Merxia	808 Merxia					28	
Misa	569 Misa					26	
Naëma	845 Naëma					7	
Nemesis	128 Nemesis					29	Concordia family after 58 Concordia
Rafita	1644 Rafita					22	
Veritas	490 Veritas					29	Undina family after 92 Undina
Theobalda	778 Theobalda	3.16 to 3.19	0.24 to 0.27	14 to 15			
***TNO families:*[D]**							
Haumea	136108 Haumea	~43	~0.19	~28			

Notes for table:

- [A]: Mean of the "core" members found in HCM and WAM analyses by Zappala et al. (1995), rounded to 2 significant digits. That analysis considered 12487 asteroids, but currently over 300,000 are known (an increase by a factor of over 25). Hence, the number of currently catalogued asteroids that are members of a given family is likely to be greater than the value in this column by a similar factor of roughly 25.
- [B]: Reference elsewhere.
- [C]: Most of these are families listed as "robustly" identified in Bendjoya and Zappala (2002). Exception: Karin family.
- [D]: TNOs are not considered asteroids, but are included here for completeness.

See also Category:Asteroid groups and families, which names some less prominent and uncertain groupings.

See also

- asteroid
- minor planet
- main belt
- Hirayama families
- proper orbital elements
- Category:Asteroid groups and families
- Haumea

References

[1] Michael E. Brown, Kristina M. Barkume, Darin Ragozzine & Emily L. Schaller, *A collisional family of icy objects in the Kuiper belt*, Nature, **446**, (March 2007), pp 294-296.

[2] David Nesvorný, Brian L. Enke, William F. Bottke, Daniel D. Durda, Erik Ashaug & Derek C. Richardson *Karin cluster formation by asteroid impact*, Icarus **183**, (2006) pp 296-311.

[3] The Hansa Family: A New High-Inclination Asteroid Family (http://adsabs.harvard.edu/abs/1996DPS....28.1007H)

- Bendjoya, Philippe; and Zappalà, Vincenzo; "Asteroid Family Identification", in *Asteroids III*, pp. 613–618, University of Arizona Press (2002), ISBN 0-8165-2281-2
- V. Zappalà et al. "Physical and Dynamical Properties of Asteroid Families", in *Asteroids III*, pp. 619–631, University of Arizona Press (2002), ISBN 0-8165-2281-2
- A. Cellino et al. "Spectroscopic Properties of Asteroid Families", in *Asteroids III*, pp. 633–643, University of Arizona Press (2002), ISBN 0-8165-2281-2
- Hirayama, Kiyotsugu; "Groups of asteroids probably of common origin", *Astronomical Journal*, Vol. 31, No. 743, pp. 185-188 (October 1918). (http://adsabs.harvard.edu/cgi-bin/nph-bib_query?bibcode=1918AJ.....31..185H&db_key=AST&high=40daf3f6f907959)
- Nesvorný, David; Bottke Jr., William F.; Dones, Luke; and Levison, Harold F.; "The recent breakup of an asteroid in the main-belt region", *Nature*, Vol. 417, pp. 720-722 (June 2002). (http://www.boulder.swri.edu/~davidn/papers/nesvorny-etal-karin-nature-2002.pdf)
- Zappalà, Vincenzo; Cellino, Alberto; Farinella, Paolo; and Knežević, Zoran; "Asteroid families I - Identification by hierarchical clustering and reliability assessment", *Astronomical Journal*, Vol. 100, p. 2030 (December 1990). (http://adsabs.harvard.edu/cgi-bin/nph-bib_query?bibcode=1990AJ....100.2030Z&db_key=AST&high=40daf3f6f915976)
- Zappalà, Vincenzo; Cellino, Alberto; Farinella, Paolo; and Milani, Andrea; "Asteroid families II - Extension to unnumbered multiopposition asteroids", *Astronomical Journal*, Vol. 107, pp. 772-801 (February 1994) (http://adsabs.harvard.edu/cgi-bin/nph-bib_query?bibcode=1994AJ....107..772Z&db_key=AST&high=40daf3f6f913642)
- V. Zappalà et al. *Asteroid Families: Search of a 12,487-Asteroid Sample Using Two Different Clustering Techniques*, Icarus, Vol. 116, p. 291 (1995.)
- M. S. Kelley & M. J. Gaffey *9 Metis and 113 Amalthea: A Genetic Asteroid Pair*, Icarus Vol. 144, p. 27 (2000).

External links

- Planetary Data System - Asteroid Families dataset (http://www.psi.edu/pds/resource/family.html), as per the Zappalà 1995 analysis.
- Latest calculations of proper elements for numbered minor planets at astDys (http://hamilton.dm.unipi.it/astdys/index.php?pc=5).
- Asteroid (and Comet) Groups (http://sajri.astronomy.cz/asteroidgroups/groups.htm) by Petr Scheirich (with excellent plots).

Solar_System

The **Solar System**[a] consists of the Sun and the astronomical objects gravitationally bound in orbit around it, all of which formed from the collapse of a giant molecular cloud approximately 4.6 billion years ago. The vast majority of the system's mass (well over 99%) is in the Sun. Of the many objects that orbit the Sun, most of the mass is contained within eight relatively solitary planets[e] whose orbits are almost circular and lie within a nearly flat disc called the ecliptic plane. The four smaller inner planets, Mercury, Venus, Earth and Mars, also called the terrestrial planets, are primarily composed of rock and metal. The four outer planets, the gas giants, are substantially more massive than the terrestrials. The two largest, Jupiter and Saturn, are composed mainly of hydrogen and helium; the two outermost planets, Uranus and Neptune, are composed largely of ices, such as water, ammonia and methane, and are often referred to separately as "ice giants".

Planets and dwarf planets of the Solar System. Sizes are to scale, but relative distances from the Sun are not.

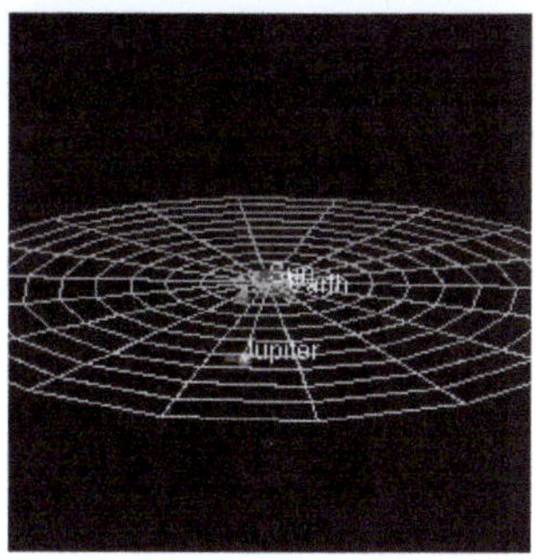
Solar System showing the plane of the ecliptic of the Earth's orbit around the Sun in 3D view showing Mercury, Venus, Earth, Mars and Jupiter making one full revolution. Saturn and Uranus also appear in their own respective orbits around the Sun

The Solar System is also home to a number of regions populated by smaller objects. The asteroid belt, which lies between Mars and Jupiter, is similar to the terrestrial planets as it is composed mainly of rock and metal. Beyond Neptune's orbit lie the Kuiper belt and scattered disc; linked populations of trans-Neptunian objects composed mostly of ices such as water, ammonia and methane. Within these populations, five individual objects, Ceres, Pluto, Haumea, Makemake and Eris, are recognized to be large enough to have been rounded by their own gravity, and are thus termed dwarf planets.[e] In addition to thousands of small bodies[e] in those two regions, various other small body populations, such as comets, centaurs and interplanetary dust, freely travel between regions.

Six of the planets and three of the dwarf planets are orbited by natural satellites,[b] usually termed "moons" after Earth's Moon. Each of the outer planets is encircled by planetary rings of dust and other particles.

The solar wind, a flow of plasma from the Sun, creates a bubble in the interstellar medium known as the heliosphere, which extends out to the edge of the scattered disc. The hypothetical Oort cloud, which acts as the source for long-period comets, may also exist at a distance roughly a thousand times further than the heliosphere.

The Solar System is located in the Milky Way galaxy, which contains about 200 billion stars.

Discovery and exploration

For many thousands of years, humanity, with a few notable exceptions, did not recognize the existence of the Solar System. People believed the Earth to be stationary at the centre of the universe and categorically different from the divine or ethereal objects that moved through the sky. Although the Greek philosopher Aristarchus of Samos had speculated on a heliocentric reordering of the cosmos,[1] Nicolaus Copernicus was the first to develop a mathematically predictive heliocentric system.[2] His 17th-century successors, Galileo Galilei, Johannes Kepler and Isaac Newton, developed an understanding of physics that led to the gradual acceptance of the idea that the Earth moves around the Sun and that the planets are governed by the same physical laws that governed the Earth. Additionally, the invention of the telescope led to the discovery of further planets and moons. In more recent times, improvements in the telescope and the use of unmanned spacecraft have enabled the investigation of geological phenomena such as mountains and craters, and seasonal meteorological phenomena such as clouds, dust storms and ice caps on the other planets.

Structure

The principal component of the Solar System is the Sun, a main-sequence G2 star that contains 99.86 percent of the system's known mass and dominates it gravitationally.[3] The Sun's four largest orbiting bodies, the gas giants, account for 99 percent of the remaining mass, with Jupiter and Saturn together comprising more than 90 percent.[c]

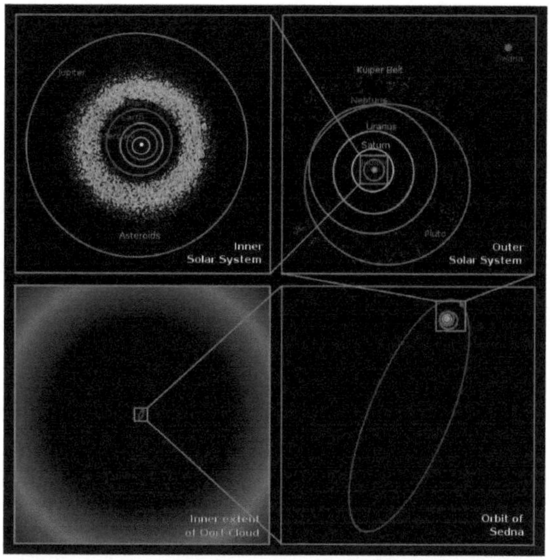

The orbits of the bodies in the Solar System to scale (clockwise from top left)

Most large objects in orbit around the Sun lie near the plane of Earth's orbit, known as the ecliptic. The planets are very close to the ecliptic while comets and Kuiper belt objects are frequently at significantly greater angles to it.[4][5] All the planets and most other objects orbit the Sun in the same direction that the Sun is rotating (counter-clockwise, as viewed from above the Sun's north pole).[6] There are exceptions, such as Halley's Comet.

The overall structure of the charted regions of the Solar System consists of the Sun, four relatively small inner planets surrounded by a belt of rocky asteroids, and four gas giants surrounded by the outer Kuiper belt of icy objects. Astronomers sometimes informally divide this structure into separate regions. The *inner Solar System* includes the four terrestrial planets and the asteroid belt. The *outer Solar System* is beyond the asteroids, including the four gas giant planets.[7] Since the discovery of the Kuiper belt, the outermost parts of the Solar System are considered a distinct region consisting of the objects beyond Neptune.[8]

Kepler's laws of planetary motion describe the orbits of objects about the Sun. Following Kepler's laws, each object travels along an ellipse with the Sun at one focus. Objects closer to the Sun (with smaller semi-major axes) travel more quickly, as they are more affected by the Sun's gravity. On an elliptical orbit, a body's distance from the Sun varies over the course of its year. A body's closest approach to the Sun is called its *perihelion*, while its most distant point from the Sun is called its *aphelion*. The orbits of the planets are nearly circular, but many comets, asteroids and Kuiper belt objects follow highly elliptical orbits.

Due to the vast distances involved, many representations of the Solar System show orbits the same distance apart. In reality, with a few exceptions, the farther a planet or belt is from the Sun, the larger the distance between it and the previous orbit. For example, Venus is approximately 0.33 astronomical units (AU)[d] farther out from the Sun than Mercury, while Saturn is 4.3 AU out from Jupiter, and Neptune lies 10.5 AU out from Uranus. Attempts have been made to determine a relationship between these orbital distances (for example, the Titius–Bode law),[9] but no such theory has been accepted.

Most of the planets in the Solar System possess secondary systems of their own, being orbited by planetary objects called natural satellites, or moons (two of which are larger than the planet Mercury), or, in the case of the four gas giants, by planetary rings; thin bands of tiny particles that orbit them in unison. Most of the largest natural satellites are in synchronous rotation, with one face permanently turned toward their parent.

Range of selected bodies of the Solar System from the middle of the Sun. The left and right edges of each bar correspond to the perihelion and aphelion of the body, respectively. Long bars denote high orbital eccentricity.

Composition

The Sun, which comprises nearly all the matter in the Solar System, is composed of roughly 98% hydrogen and helium.[10] Jupiter and Saturn, which comprise nearly all the remaining matter, possess atmospheres composed of roughly 99% of those same elements.[11] [12] A composition gradient exists in the Solar System, created by heat and light pressure from the Sun; those objects closer to the Sun, which are more affected by heat and light pressure, are composed of elements with high melting points. Objects farther from the Sun are composed largely of materials with lower melting points.[13] The boundary in the Solar System beyond which those volatile substances could condense is known as the frost line, and it lies at roughly 4 AU from the Sun.[14]

The objects of the inner Solar System are composed mostly of *rock*,[15] the collective name for compounds with high melting points, such as silicates, iron or nickel, that remained solid under almost all conditions in the protoplanetary nebula.[16] Jupiter and Saturn are composed mainly of *gases*, the astronomical term for materials with extremely low melting points and high vapor pressure such as molecular hydrogen, helium, and neon, which were always in the gaseous phase in the nebula.[16] *Ices*, like water, methane, ammonia, hydrogen sulfide and carbon dioxide,[15] have melting points up to a few hundred kelvins, while their phase depends on the ambient pressure and temperature.[16] They can be found as ices, liquids, or gases in various places in the Solar System, while in the nebula they were either in the solid or gaseous phase.[16] Icy substances comprise the majority of the satellites of the giant planets, as well as most of Uranus and Neptune (the so-called "ice giants") and the numerous small objects that lie beyond Neptune's orbit.[15] [17] Together, gases and ices are referred to as *volatiles*.[18]

Sun

The Sun is the Solar System's star, and by far its chief component. Its large mass (332,900 Earth masses)[19] produces temperatures and densities in its core great enough to sustain nuclear fusion,[20] which releases enormous amounts of energy, mostly radiated into space as electromagnetic radiation, peaking in the 400–700 nm band we call visible light.[21]

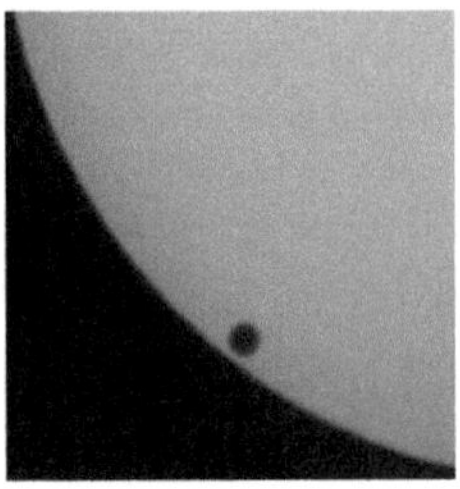
A transit of Venus

The Sun is classified as a type G2 yellow dwarf, but this name is misleading as, compared to the majority of stars in our galaxy, the Sun is rather large and bright.[22] Stars are classified by the Hertzsprung–Russell diagram, a graph that plots the brightness of stars with their surface temperatures. Generally, hotter stars are brighter. Stars following this pattern are said to be on the main sequence, and the Sun lies right in the middle of it. However, stars brighter and hotter than the Sun are rare, while substantially dimmer and cooler stars, known as red dwarfs, are common, making up 85 percent of the stars in the galaxy.[22] [23]

Evidence suggests that the Sun's position on the main sequence puts it in the "prime of life" for a star, in that it has not yet exhausted its store of hydrogen for nuclear fusion. The Sun is growing brighter; early in its history it was 70 percent as bright as it is today.[24]

The Sun is a population I star; it was born in the later stages of the universe's evolution, and thus contains more elements heavier than hydrogen and helium ("metals" in astronomical parlance) than older population II stars.[25] Elements heavier than hydrogen and helium were formed in the cores of ancient and exploding stars, so the first generation of stars had to die before the universe could be enriched with these atoms. The oldest stars contain few metals, while stars born later have more. This high metallicity is thought to have been crucial to the Sun's developing a planetary system, because planets form from accretion of "metals".[26]

The heliospheric current sheet

Interplanetary medium

Along with light, the Sun radiates a continuous stream of charged particles (a plasma) known as the solar wind. This stream of particles spreads outwards at roughly 1.5 million kilometres per hour,[27] creating a tenuous atmosphere (the heliosphere) that permeates the Solar System out to at least 100 AU (see heliopause).[28] This is known as the interplanetary medium. Activity on the Sun's surface, such as solar flares and coronal mass ejections, disturb the heliosphere, creating space weather and causing geomagnetic storms.[29] The largest structure within the heliosphere is the heliospheric current sheet, a spiral form created by the actions of the Sun's rotating magnetic field on the interplanetary medium.[30] [31]

Earth's magnetic field stops its atmosphere from being stripped away by the solar wind. Venus and Mars do not have magnetic fields, and as a result, the solar wind causes their atmospheres to gradually bleed away into space.[32] Coronal mass ejections and similar events blow a magnetic field and huge quantities of material from the surface of the Sun. The interaction of this magnetic field and material with Earth's magnetic field funnels charged particles into the Earth's upper atmosphere, where its interactions create aurorae seen near the magnetic poles.

Cosmic rays originate outside the Solar System. The heliosphere partially shields the Solar System, and planetary magnetic fields (for those planets that have them) also provide some protection. The density of cosmic rays in the interstellar medium and the strength of the Sun's magnetic field change on very long timescales, so the level of cosmic radiation in the Solar System varies, though by how much is unknown.[33]

The interplanetary medium is home to at least two disc-like regions of cosmic dust. The first, the zodiacal dust cloud, lies in the inner Solar System and causes zodiacal light. It was likely formed by collisions within the asteroid belt brought on by interactions with the planets.[34] The second extends from about 10 AU to about 40 AU, and was probably created by similar collisions within the Kuiper belt.[35] [36]

Inner Solar System

The inner Solar System is the traditional name for the region comprising the terrestrial planets and asteroids.[37] Composed mainly of silicates and metals, the objects of the inner Solar System are relatively close to the Sun; the radius of this entire region is shorter than the distance between Jupiter and Saturn.

Inner planets

The four inner or terrestrial planets have dense, rocky compositions, few or no moons, and no ring systems. They are composed largely of refractory minerals, such as the silicates, which form their crusts and mantles, and metals such as iron and nickel, which form their cores. Three of the four inner planets (Venus, Earth and Mars) have atmospheres substantial enough to generate weather; all have impact craters and tectonic surface features such as rift valleys and volcanoes. The term *inner planet* should not be confused with *inferior planet*, which designates those planets that are closer to the Sun than Earth is (i.e. Mercury and Venus).

The inner planets. From left to right: Mercury, Venus, Earth, and Mars (sizes to scale, interplanetary distances not)

Mercury

Mercury (0.4 AU from the Sun) is the closest planet to the Sun and the smallest planet in the Solar System (0.055 Earth masses). Mercury has no natural satellites, and its only known geological features besides impact craters are lobed ridges or rupes, probably produced by a period of contraction early in its history.[38] Mercury's almost negligible atmosphere consists of atoms blasted off its surface by the solar wind.[39] Its relatively large iron core and thin mantle have not yet been adequately explained. Hypotheses include that its outer layers were stripped off by a giant impact, and that it was prevented from fully accreting by the young Sun's energy.[40] [41]

Venus

Venus (0.7 AU from the Sun) is close in size to Earth (0.815 Earth masses), and, like Earth, has a thick silicate mantle around an iron core, a substantial atmosphere and evidence of internal geological activity. However, it is much drier than Earth and its atmosphere is ninety times as dense. Venus has no natural satellites. It is the hottest planet, with surface temperatures over 400 °C, most likely due to the amount of greenhouse gases in the atmosphere.[42] No definitive evidence of current geological activity has been detected on Venus, but it has no magnetic field that would prevent depletion of its substantial atmosphere, which suggests that its atmosphere is regularly replenished by volcanic eruptions.[43]

Earth

Earth (1 AU from the Sun) is the largest and densest of the inner planets, the only one known to have current geological activity, and is the only place in the Solar System where life is known to exist.[44] Its liquid hydrosphere is unique among the terrestrial planets, and it is also the only planet where plate tectonics has been observed. Earth's atmosphere is radically different from those of the other planets, having been altered by the presence of life to contain 21% free oxygen.[45] It has one natural satellite, the Moon, the only large satellite of a terrestrial planet in the Solar System.

Mars

Mars (1.5 AU from the Sun) is smaller than Earth and Venus (0.107 Earth masses). It possesses an atmosphere of mostly carbon dioxide with a surface pressure of 6.1 millibars (roughly 0.6 percent that of the Earth's).[46] Its surface, peppered with vast volcanoes such as Olympus Mons and rift valleys such as Valles Marineris, shows geological activity that may have persisted until as recently as 2 million years ago.[47] Its red colour comes from iron oxide (rust) in its soil.[48] Mars has two tiny natural satellites (Deimos and Phobos) thought to be captured asteroids.[49]

Asteroid belt

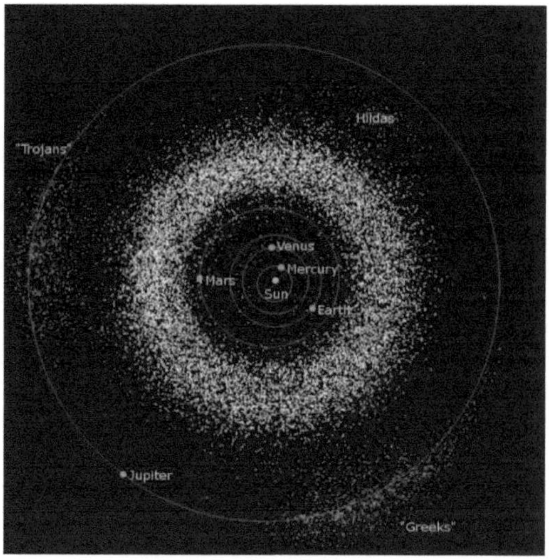

Image of the asteroid belt and the Trojan asteroids

Asteroids are mostly small Solar System bodies[e] composed mainly of refractory rocky and metallic minerals, with some ice.[50]

The asteroid belt occupies the orbit between Mars and Jupiter, between 2.3 and 3.3 AU from the Sun. It is thought to be remnants from the Solar System's formation that failed to coalesce because of the gravitational interference of Jupiter.[51]

Asteroids range in size from hundreds of kilometres across to microscopic. All asteroids except the largest, Ceres, are classified as small Solar System bodies, but some asteroids such as Vesta and Hygiea may be reclassed as dwarf planets if they are shown to have achieved hydrostatic equilibrium.[52]

The asteroid belt contains tens of thousands, possibly millions, of objects over one kilometre in diameter.[53] Despite this, the total mass of the asteroid belt is unlikely to be more than a thousandth of that of the Earth.[54] The asteroid belt is very sparsely populated; spacecraft routinely pass through without incident. Asteroids with diameters between 10 and 10^{-4} m are called meteoroids.[55]

Ceres

Ceres (2.77 AU) is the largest asteroid, a protoplanet, and a dwarf planet.[e] It has a diameter of slightly under 1000 km, and a mass large enough for its own gravity to pull it into a spherical shape. Ceres was considered a planet when it was discovered in the 19th century, but was reclassified as an asteroid in the 1850s as further observations revealed additional asteroids.[56] It was classified in 2006 as a dwarf planet.

Asteroid groups

Asteroids in the asteroid belt are divided into asteroid groups and families based on their orbital characteristics. Asteroid moons are asteroids that orbit larger asteroids. They are not as clearly distinguished as planetary moons, sometimes being almost as large as their partners. The asteroid belt also contains main-belt comets, which may have been the source of Earth's water.[57]

Trojan asteroids are located in either of Jupiter's L_4 or L_5 points (gravitationally stable regions leading and trailing a planet in its orbit); the term "Trojan" is also used for small bodies in any other planetary or satellite Lagrange point. Hilda asteroids are in a 2:3 resonance with Jupiter; that is, they go around the Sun three times for every two Jupiter orbits.[58]

The inner Solar System is also dusted with rogue asteroids, many of which cross the orbits of the inner planets.[59]

Outer Solar System

The outer region of the Solar System is home to the gas giants and their large moons. Many short-period comets, including the centaurs, also orbit in this region. Due to their greater distance from the Sun, the solid objects in the outer Solar System contain a higher proportion of volatiles such as water, ammonia and methane, than the rocky denizens of the inner Solar System, as the colder temperatures allow these compounds to remain solid.

Outer planets

The four outer planets, or gas giants (sometimes called Jovian planets), collectively make up 99 percent of the mass known to orbit the Sun.[c] Jupiter and Saturn are each many tens of times the mass of the Earth and consist overwhelmingly of hydrogen and helium; Uranus and Neptune are far less massive (<20 Earth masses) and possess more ices in their makeup. For these reasons, some astronomers suggest they belong in their own category, "ice giants."[60] All four gas giants have rings, although only Saturn's ring system is easily observed from Earth. The term *outer planet* should not be confused with *superior planet*, which designates planets outside Earth's orbit and thus includes both the outer planets and Mars.

From top to bottom: Neptune, Uranus, Saturn, and Jupiter (not to scale)

Jupiter

Jupiter (5.2 AU), at 318 Earth masses, is 2.5 times the mass of all the other planets put together. It is composed largely of hydrogen and helium. Jupiter's strong internal heat creates a number of semi-permanent features in its atmosphere, such as cloud bands and the Great Red Spot.

Jupiter has 64 known satellites. The four largest, Ganymede, Callisto, Io, and Europa, show similarities to the terrestrial planets, such as volcanism and internal heating.[61] Ganymede, the largest satellite in the Solar System, is larger than Mercury.

Saturn

Saturn (9.5 AU), distinguished by its extensive ring system, has several similarities to Jupiter, such as its atmospheric composition and magnetosphere. Although Saturn has 60% of Jupiter's volume, it is less than a third as massive, at 95 Earth masses, making it the least dense planet in the Solar System. The rings of Saturn are made up of small ice and rock particles.

Saturn has 62 confirmed satellites; two of which, Titan and Enceladus, show signs of geological activity, though they are largely made of ice.[62] Titan, the second-largest moon in the Solar System, is larger than Mercury and the only satellite in the Solar System with a substantial atmosphere.

Uranus

Uranus (19.6 AU), at 14 Earth masses, is the lightest of the outer planets. Uniquely among the planets, it orbits the Sun on its side; its axial tilt is over ninety degrees to the ecliptic. It has a much colder core than the other gas giants, and radiates very little heat into space.[63]

Uranus has 27 known satellites, the largest ones being Titania, Oberon, Umbriel, Ariel and Miranda.

Neptune

Neptune (30 AU), though slightly smaller than Uranus, is more massive (equivalent to 17 Earths) and therefore more dense. It radiates more internal heat, but not as much as Jupiter or Saturn.[64]

Neptune has 13 known satellites. The largest, Triton, is geologically active, with geysers of liquid nitrogen.[65] Triton is the only large satellite with a retrograde orbit. Neptune is accompanied in its orbit by a number of minor planets, termed Neptune Trojans, that are in 1:1 resonance with it.

Comets

Comets are small Solar System bodies,[e] typically only a few kilometres across, composed largely of volatile ices. They have highly eccentric orbits, generally a perihelion within the orbits of the inner planets and an aphelion far beyond Pluto. When a comet enters the inner Solar System, its proximity to the Sun causes its icy surface to sublimate and ionise, creating a coma: a long tail of gas and dust often visible to the naked eye.

Comet Hale–Bopp

Short-period comets have orbits lasting less than two hundred years. Long-period comets have orbits lasting thousands of years. Short-period comets are believed to originate in the Kuiper belt, while long-period comets, such as Hale–Bopp, are believed to originate in the Oort cloud. Many comet groups, such as the Kreutz Sungrazers, formed from the breakup of a single parent.[66] Some comets with hyperbolic orbits may originate outside the Solar System, but determining their precise orbits is difficult.[67] Old comets that have had most of their volatiles driven out by solar warming are often categorised as asteroids.[68]

Centaurs

The centaurs are icy comet-like bodies with a semi-major axis greater than Jupiter's (5.5 AU) and less than Neptune's (30 AU). The largest known centaur, 10199 Chariklo, has a diameter of about 250 km.[69] The first centaur discovered, 2060 Chiron, has also been classified as comet (95P) since it develops a coma just as comets do when they approach the Sun.[70]

Trans-Neptunian region

The area beyond Neptune, or the "trans-Neptunian region", is still largely unexplored. It appears to consist overwhelmingly of small worlds (the largest having a diameter only a fifth that of the Earth and a mass far smaller than that of the Moon) composed mainly of rock and ice. This region is sometimes known as the "outer Solar System", though others use that term to mean the region beyond the asteroid belt.

Kuiper belt

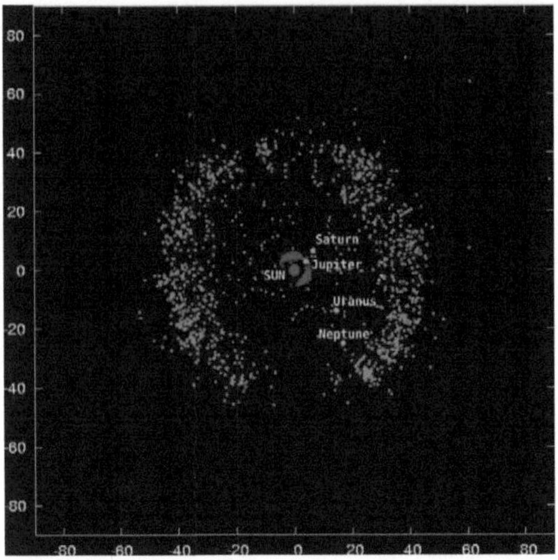

Plot of all known Kuiper belt objects, set against the four outer planets

The Kuiper belt, the region's first formation, is a great ring of debris similar to the asteroid belt, but composed mainly of ice.[71] It extends between 30 and 50 AU from the Sun. Though it contains at least three dwarf planets, it is composed mainly of small Solar System bodies. However, many of the largest Kuiper belt objects, such as Quaoar, Varuna, and Orcus, may be reclassified as dwarf planets. There are estimated to be over 100,000 Kuiper belt objects with a diameter greater than 50 km, but the total mass of the Kuiper belt is thought to be only a tenth or even a hundredth the mass of the Earth.[72] Many Kuiper belt objects have multiple satellites,[73] and most have orbits that take them outside the plane of the ecliptic.[74]

The Kuiper belt can be roughly divided into the "classical" belt and the resonances.[71] Resonances are orbits linked to that of Neptune (e.g. twice for every three Neptune orbits, or once for every two). The first resonance begins within the orbit of Neptune itself. The classical belt consists of objects having no resonance with Neptune, and extends from roughly 39.4 AU to 47.7 AU.[75] Members of the classical Kuiper belt are classified as cubewanos, after the first of their kind to be discovered, (15760) 1992 QB_1, and are still in near primordial, low-eccentricity orbits.[76]

Pluto and Charon

Pluto (39 AU average), a dwarf planet, is the largest known object in the Kuiper belt. When discovered in 1930, it was considered to be the ninth planet; this changed in 2006 with the adoption of a formal definition of planet. Pluto has a relatively eccentric orbit inclined 17 degrees to the ecliptic plane and ranging from 29.7 AU from the Sun at perihelion (within the orbit of Neptune) to 49.5 AU at aphelion.

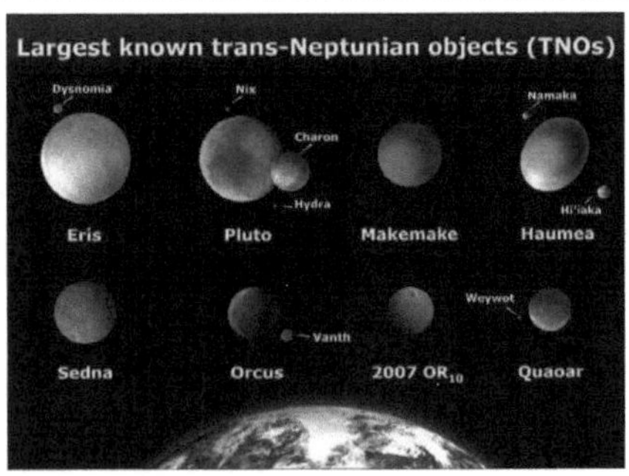

Artistic comparison of Eris, Pluto, Makemake, Haumea, Sedna, Orcus, 2007 OR10, Quaoar, and Earth (scales are outdated)

Charon, Pluto's largest moon, is sometimes described as part of a binary system with Pluto, as the two bodies orbit a barycenter of gravity above their surfaces (i.e., they appear to "orbit each other"). Beyond Charon, three much smaller moons, Nix, P4 and Hydra, orbit within the system.

Pluto has a 3:2 resonance with Neptune, meaning that Pluto orbits twice round the Sun for every three Neptunian orbits. Kuiper belt objects whose orbits share this resonance are called plutinos.[77]

Haumea and Makemake

Haumea (43.34 AU average), and Makemake (45.79 AU average), while smaller than Pluto, are the largest known objects in the *classical* Kuiper belt (that is, they are not in a confirmed resonance with Neptune). Haumea is an egg-shaped object with two moons. Makemake is the brightest object in the Kuiper belt after Pluto. Originally designated **2003 EL$_{61}$** and **2005 FY$_{9}$** respectively, they were given names and designated dwarf planets in 2008.[78] Their orbits are far more inclined than Pluto's, at 28° and 29°.[79]

Scattered disc

The scattered disc, which overlaps the Kuiper belt but extends much further outwards, is thought to be the source of short-period comets. Scattered disc objects are believed to have been ejected into erratic orbits by the gravitational influence of Neptune's early outward migration. Most scattered disc objects (SDOs) have perihelia within the Kuiper belt but aphelia as far as 150 AU from the Sun. SDOs' orbits are also highly inclined to the ecliptic plane, and are often almost perpendicular to it. Some astronomers consider the scattered disc to be merely another region of the Kuiper belt, and describe scattered disc objects as "scattered Kuiper belt objects."[80] Some astronomers also classify centaurs as inward-scattered Kuiper belt objects along with the outward-scattered residents of the scattered disc.[81]

Eris

Eris (68 AU average) is the largest known scattered disc object, and caused a debate about what constitutes a planet, since it is 25% more massive than Pluto[82] and about the same diameter. It is the most massive of the known dwarf planets. It has one moon, Dysnomia. Like Pluto, its orbit is highly eccentric, with a perihelion of 38.2 AU (roughly Pluto's distance from the Sun) and an aphelion of 97.6 AU, and steeply inclined to the ecliptic plane.

Farthest regions

The point at which the Solar System ends and interstellar space begins is not precisely defined, since its outer boundaries are shaped by two separate forces: the solar wind and the Sun's gravity. The outer limit of the solar wind's influence is roughly four times Pluto's distance from the Sun; this *heliopause* is considered the beginning of the interstellar medium.[28] However, the Sun's Roche sphere, the effective range of its gravitational dominance, is believed to extend up to a thousand times farther.[83]

Heliopause

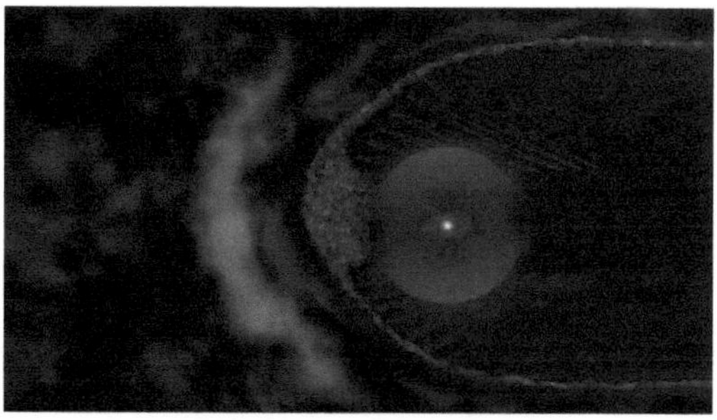

NASA image of the heliosheath and heliopause

The heliosphere is divided into two separate regions. The solar wind travels at roughly 400 km/s until it collides with the interstellar wind; the flow of plasma in the interstellar medium. The collision occurs at the termination shock, which is roughly 80–100 AU from the Sun upwind of the interstellar medium and roughly 200 AU from the Sun downwind.[84] Here the wind slows dramatically, condenses and becomes more turbulent,[84] forming a great oval structure known as the heliosheath. This structure is believed to look and behave very much like a comet's tail, extending outward for a further 40 AU on the upwind side but tailing many times that distance downwind; but evidence from the Cassini and Interstellar Boundary Explorer spacecraft has suggested that it is in fact forced into a bubble shape by the constraining action of the interstellar magnetic field.[85] Both *Voyager 1* and *Voyager 2* are reported to have passed the termination shock and entered the heliosheath, at 94 and 84 AU from the Sun, respectively.[86] [87] The outer boundary of the heliosphere, the heliopause, is the point at which the solar wind finally terminates and is the beginning of interstellar space.[28]

The shape and form of the outer edge of the heliosphere is likely affected by the fluid dynamics of interactions with the interstellar medium[84] as well as solar magnetic fields prevailing to the south, e.g. it is bluntly shaped with the northern hemisphere extending 9 AU farther than the southern hemisphere. Beyond the heliopause, at around 230 AU, lies the bow shock, a plasma "wake" left by the Sun as it travels through the Milky Way.[88]

No spacecraft have yet passed beyond the heliopause, so it is impossible to know for certain the conditions in local interstellar space. It is expected that NASA's Voyager spacecraft will pass the heliopause some time in the next decade and transmit valuable data on radiation levels and solar wind back to the Earth.[89] How well the heliosphere shields the Solar System from cosmic rays is poorly understood. A NASA-funded team has developed a concept of a "Vision Mission" dedicated to sending a probe to the heliosphere.[90] [91]

Oort cloud

The hypothetical Oort cloud is a spherical cloud of up to a trillion icy objects that is believed to be the source for all long-period comets and to surround the Solar System at roughly 50,000 AU (around 1 light-year (LY)), and possibly to as far as 100,000 AU (1.87 LY). It is believed to be composed of comets that were ejected from the inner Solar System by gravitational interactions with the outer planets. Oort cloud objects move very slowly, and can be perturbed by infrequent events such as collisions, the gravitational effects of a passing star, or the galactic tide, the tidal force exerted by the Milky Way.[92] [93]

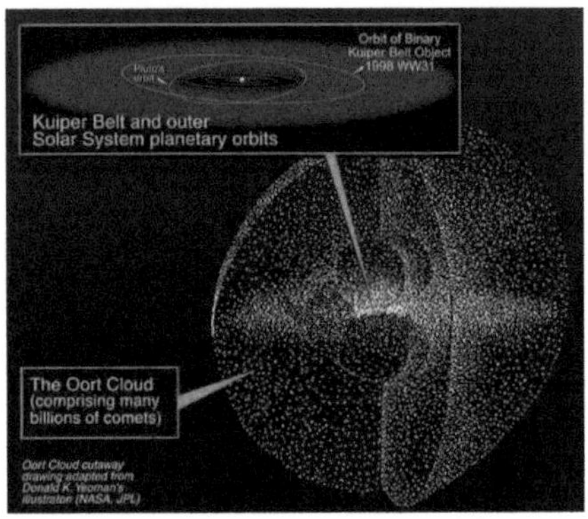

An artist's rendering of the Oort Cloud, the Hills Cloud, and the Kuiper belt (inset)

Sedna

90377 Sedna (525.86 AU average) is a large, reddish Pluto-like object with a gigantic, highly elliptical orbit that takes it from about 76 AU at perihelion to 928 AU at aphelion and takes 12,050 years to complete. Mike Brown, who discovered the object in 2003, asserts that it cannot be part of the scattered disc or the Kuiper belt as its perihelion is too distant to have been affected by Neptune's migration. He and other astronomers consider it to be the first in an entirely new population, which also may include the object 2000 CR_{105}, which has a perihelion of 45 AU, an aphelion of 415 AU, and an orbital period of 3,420 years.[94] Brown terms this population the "Inner Oort cloud," as it may have formed through a similar process, although it is far closer to the Sun.[95] Sedna is very likely a dwarf planet, though its shape has yet to be determined with certainty.

Boundaries

Much of our Solar System is still unknown. The Sun's gravitational field is estimated to dominate the gravitational forces of surrounding stars out to about two light years (125,000 AU). Lower estimates for the radius of the Oort cloud, by contrast, do not place it farther than 50,000 AU.[96] Despite discoveries such as Sedna, the region between the Kuiper belt and the Oort cloud, an area tens of thousands of AU in radius, is still virtually unmapped. There are also ongoing studies of the region between Mercury and the Sun.[97] Objects may yet be discovered in the Solar System's uncharted regions.

Galactic context

The Solar System is located in the Milky Way galaxy, a barred spiral galaxy with a diameter of about 100,000 light-years containing about 200 billion stars.[98] Our Sun resides in one of the Milky Way's outer spiral arms, known as the Orion Arm or Local Spur.[99] The Sun lies between 25,000 and 28,000 light years from the Galactic Centre,[100] and its speed within the galaxy is about 220 kilometres per second, so that it completes one revolution every 225–250 million years. This revolution is known as the Solar System's galactic year.[101] The solar apex, the direction of the Sun's path through interstellar space, is near the constellation of Hercules in the direction of the current location of the bright star Vega.[102] The plane of the ecliptic lies at an angle of about 60° to the galactic plane.[f]

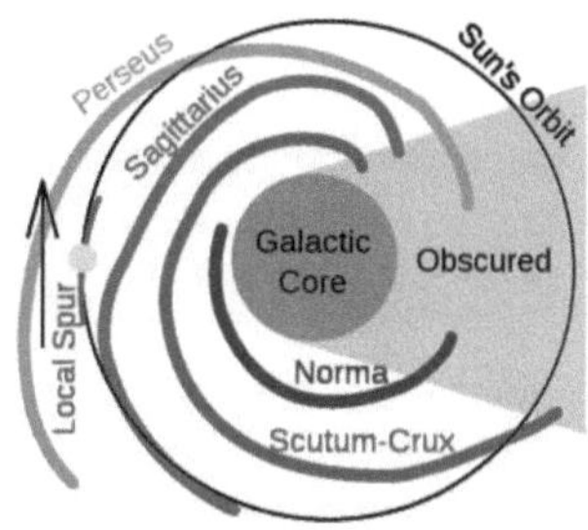

Location of the Solar System within our galaxy

The Solar System's location in the galaxy is very likely a factor in the evolution of life on Earth. Its orbit is close to being circular and is at roughly the same speed as that of the spiral arms, which means it passes through them only rarely. Since spiral arms are home to a far larger concentration of potentially dangerous supernovae, this has given Earth long periods of interstellar stability for life to evolve.[103] The Solar System also lies well outside the star-crowded environs of the galactic centre. Near the centre, gravitational tugs from nearby stars could perturb bodies in the Oort Cloud and send many comets into the inner Solar System, producing collisions with potentially catastrophic implications for life on Earth. The intense radiation of the galactic centre could also interfere with the development of complex life.[103] Even at the Solar System's current location, some scientists have hypothesised that recent supernovae may have adversely affected life in the last 35,000 years by flinging pieces of expelled stellar core towards the Sun as radioactive dust grains and larger, comet-like bodies.[104]

Neighbourhood

The immediate galactic neighbourhood of the Solar System is known as the Local Interstellar Cloud or Local Fluff, an area of denser cloud in an otherwise sparse region known as the Local Bubble, an hourglass-shaped cavity in the interstellar medium roughly 300 light years across. The bubble is suffused with high-temperature plasma that suggests it is the product of several recent supernovae.[105]

There are relatively few stars within ten light years (95 trillion km) of the Sun. The closest is the triple star system Alpha Centauri, which is about 4.4 light years away. Alpha Centauri A and B are a closely tied pair of Sun-like stars, while the small red dwarf Alpha Centauri C (also known as Proxima Centauri) orbits the pair at a distance of 0.2 light years. The stars next closest to the Sun are the red dwarfs Barnard's Star (at 5.9 light years), Wolf 359 (7.8 light years) and Lalande 21185 (8.3 light years). The largest star within ten light years is Sirius, a bright main-sequence star roughly twice the Sun's mass and orbited by a white dwarf called Sirius B. It lies 8.6 light years away. The remaining systems within ten light years are the binary red dwarf system Luyten 726-8 (8.7 light years) and the solitary red dwarf Ross 154 (9.7 light years).[106] Our closest solitary sun-like star is Tau Ceti, which lies 11.9 light years away. It has roughly 80 percent the Sun's mass, but only 60 percent its luminosity.[107] The closest known extrasolar planet to the Sun lies around the star Epsilon Eridani, a star slightly dimmer and redder than the Sun, which lies 10.5 light years away. Its one confirmed planet, Epsilon Eridani b, is roughly 1.5 times Jupiter's mass and orbits its star every 6.9 years.[108]

A diagram of our location in the observable Universe. (*Click here for larger image* [109].)

Formation and evolution

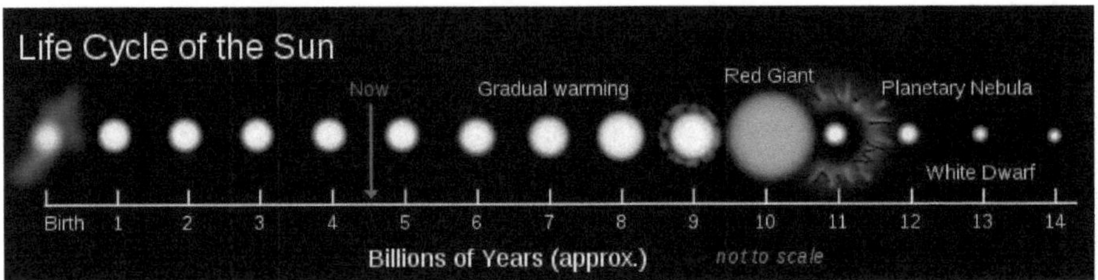

The Solar System formed from the gravitational collapse of a giant molecular cloud 4.568 billion years ago.[110] This initial cloud was likely several light-years across and probably birthed several stars.[111]

As the region that would become the Solar System, known as the pre-solar nebula,[112] collapsed, conservation of angular momentum made it rotate faster. The centre, where most of the mass collected, became increasingly hotter than the surrounding disc.[111] As the contracting nebula rotated, it began to flatten into a spinning protoplanetary disc with a diameter of roughly 200 AU[111] and a hot, dense protostar at the centre.[113] [114] At this point in its evolution, the Sun is believed to have been a T Tauri star. Studies of T Tauri stars show that they are often accompanied by discs of pre-planetary matter with masses of 0.001–0.1 solar masses, with the vast majority of the mass of the nebula in the star itself.[115] The planets formed by accretion from this disk.[116]

Within 50 million years, the pressure and density of hydrogen in the centre of the protostar became great enough for it to begin thermonuclear fusion.[117] The temperature, reaction rate, pressure, and density increased until hydrostatic equilibrium was achieved, with the thermal energy countering the force of gravitational contraction. At this point the Sun became a full-fledged main-sequence star.[118]

The Solar System as we know it today will last until the Sun begins its evolution off the main sequence of the Hertzsprung–Russell diagram. As the Sun burns through its supply of hydrogen fuel, the energy output supporting the core tends to decrease, causing it to collapse in on itself. This increase in pressure heats the core, so it burns even faster. As a result, the Sun is growing brighter at a rate of roughly ten percent every 1.1 billion years.[119]

Around 5.4 billion years from now, the hydrogen in the core of the Sun will have been entirely converted to helium, ending the main-sequence phase. As the hydrogen reactions shut down, the core will contract further, increasing pressure and temperature, causing fusion to commence via the helium process. Helium in the core burns at a much hotter temperature, and the energy output will be much greater than during the hydrogen process. At this time, the outer layers of the Sun will expand to roughly up to 260 times its current diameter; the Sun will become a red giant. Because of its vastly increased surface area, the surface of the Sun will be considerably cooler than it is on the main sequence (2600 K at the coolest).[120]

Eventually, helium in the core will exhaust itself at a much faster rate than the hydrogen, and the Sun's helium burning phase will be but a fraction of the time compared to the hydrogen burning phase. The Sun is not massive enough to commence fusion of heavier elements, and nuclear reactions in the core will dwindle. Its outer layers will fall away into space, leaving a white dwarf, an extraordinarily dense object, half the original mass of the Sun but only the size of the Earth.[121] The ejected outer layers will form what is known as a planetary nebula, returning some of the material that formed the Sun to the interstellar medium.

Visual summary

A sampling of closely imaged Solar System bodies, selected for size and detail. The Sun is approximately 10,000 times larger than, and 41 trillion times the volume of, the smallest object shown (Prometheus). See also List of Solar System objects by size, List of natural satellites, List of minor planets, and Lists of comets. Images here are not an endorsement of natural color in visible light.

Solar System Summary

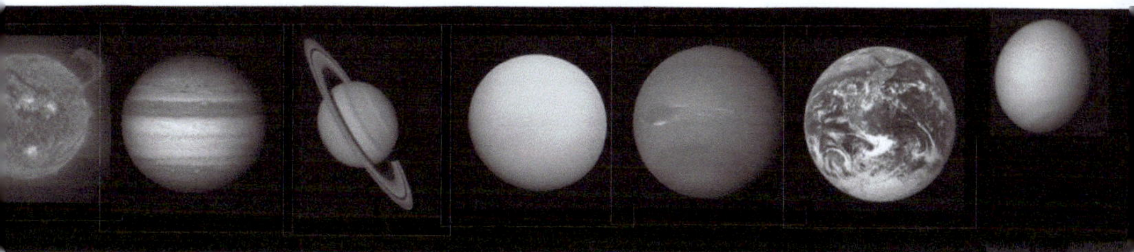

Sun	Jupiter	Saturn	Uranus	Neptune	Earth	Venus

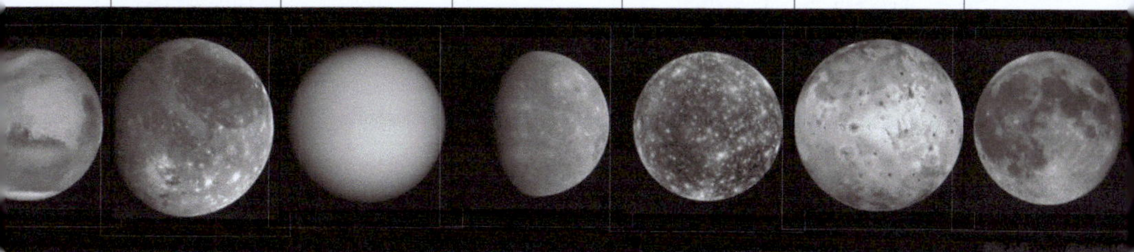

Mars	Ganymede	Titan	Mercury	Callisto	Io	Moon

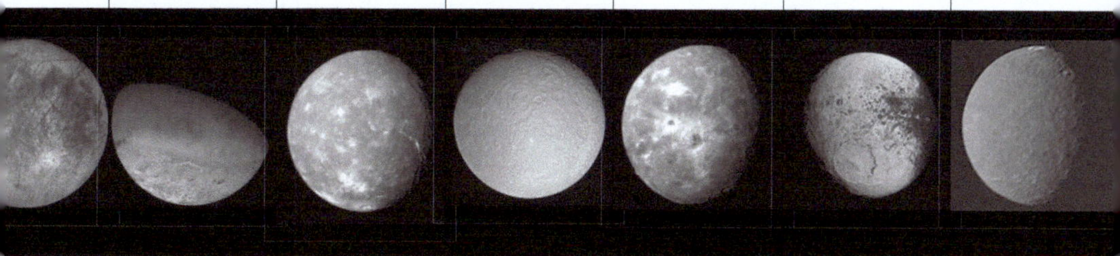

Europa	Triton	Titania	Rhea	Oberon	Iapetus	Umbriel

Ariel	Dione	Tethys	Vesta	Enceladus	Miranda	Proteus

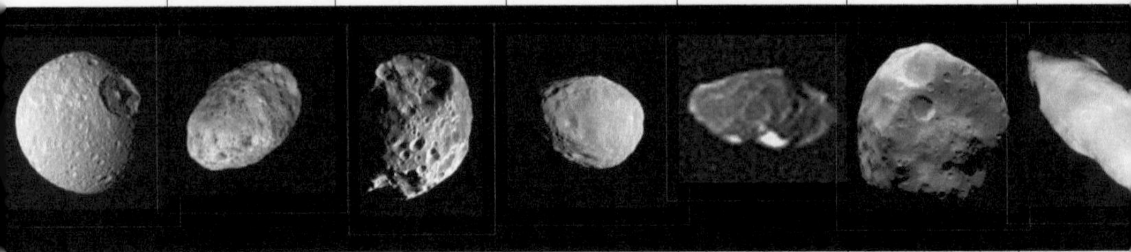

Mimas	Hyperion	Phoebe	Janus	Amalthea	Epimetheus	Promethe

See also

- Astronomical symbols
- List of geological features of the Solar System
- Numerical model of the Solar System
- Orrery, mechanical models of Solar System
- Planetary mnemonic
- Solar System in fiction
- Solar System model

Notes

a. Capitalization of the name varies. The IAU, the authoritative body regarding astronomical nomenclature, specifies capitalizing the names of all individual astronomical objects [122] (**Solar System**). However, the name is commonly rendered in lower case (**solar system**) – as, for example, in the *Oxford English Dictionary* and *Merriam-Webster's 11th Collegiate Dictionary* [123]

b. See List of natural satellites for the full list of natural satellites of the eight planets and five dwarf planets.

c. The mass of the Solar System excluding the Sun, Jupiter and Saturn can be determined by adding together all the calculated masses for its largest objects and using rough calculations for the masses of the Oort cloud (estimated at roughly 3 Earth masses),[124] the Kuiper belt (estimated at roughly 0.1 Earth mass)[72] and the asteroid belt (estimated to be 0.0005 Earth mass)[54] for a total, rounded upwards, of ~37 Earth masses, or 8.1 percent the mass in orbit around the Sun. With the combined masses of Uranus and Neptune (~31 Earth masses) subtracted, the remaining ~6 Earth masses of material comprise 1.3 percent of the total.

d. Astronomers measure distances within the Solar System in astronomical units (AU). One AU equals the average distance between the centers of Earth and the Sun, or 149,598,000 km. Pluto is about 38 AU from the Sun and Jupiter is about 5.2 AU from the Sun. One light-year is 63,240 AU.

e. According to current definitions, objects in orbit around the Sun are classed dynamically and physically into three categories: *planets*, *dwarf planets* and *small Solar System bodies*. A planet is any body in orbit around the Sun that has enough mass to form itself into a spherical shape and has cleared its immediate neighbourhood of all smaller objects. By this definition, the Solar System has eight known planets: Mercury, Venus, Earth, Mars, Jupiter, Saturn, Uranus, and Neptune. Pluto does not fit this definition, as it has not cleared its orbit of surrounding Kuiper belt objects.[125] A dwarf planet is a celestial body orbiting the Sun that is massive enough to be rounded by its own gravity but has not cleared its neighbouring region of planetesimals and is not a satellite.[125] By this definition, the Solar System has five known dwarf planets: Ceres, Pluto, Haumea, Makemake, and Eris.[78] Other objects may be classified in the future as dwarf planets, such as Sedna, Orcus, and Quaoar.[126] Dwarf planets that orbit in the trans-Neptunian region are called "plutoids".[127] The remainder of the objects in orbit around the Sun are small Solar System bodies.[125]

f. If ψ is the angle between the north pole of the ecliptic and the north galactic pole then:

,

where 27° 07′ 42.01″ and 12h 51m 26.282 are the declination and right ascension of the north galactic pole,[128] while 66° 33′ 38.6″ and 18h 0m 00 are those for the north pole of the ecliptic. (Both pairs of coordinates are for J2000 epoch.) The result of the calculation is 60.19°.

References

[1] WC Rufus (1923). "The astronomical system of Copernicus". *Popular Astronomy* **31**: 510. Bibcode 1923PA.....31..510R.

[2] Weinert, Friedel (2009). *Copernicus, Darwin, & Freud: revolutions in the history and philosophy of science*. Wiley-Blackwell. p. 21. ISBN 9781405181839.

[3] M Woolfson (2000). "The origin and evolution of the solar system". *Astronomy & Geophysics* **41** (1): 1.12. doi:10.1046/j.1468-4004.2000.00012.x.

[4] Harold F. Levison, Alessandro Morbidelli (2003). "The formation of the Kuiper belt by the outward transport of bodies during Neptune's migration" (http://www.obs-nice.fr/morby/stuff/NATURE.pdf) (PDF). . Retrieved 2007-06-25.

[5] Harold F. Levison, Martin J Duncan (1997). "From the Kuiper Belt to Jupiter-Family Comets: The Spatial Distribution of Ecliptic Comets". *Icarus* **127** (1): 13–32. Bibcode 1997Icar..127...13L. doi:10.1006/icar.1996.5637.

[6] Grossman, Lisa (13 August 2009). "Planet found orbiting its star backwards for first time" (http://www.newscientist.com/article/dn17603-planet-found-orbiting-its-star-backwards-for-first-time.html). NewScientist. . Retrieved 10 October 2009.

[7] nineplanets.org. "An Overview of the Solar System" (http://www.nineplanets.org/overview.html). . Retrieved 2007-02-15.

[8] Amir Alexander (2006). "New Horizons Set to Launch on 9-Year Voyage to Pluto and the Kuiper Belt" (http://www.planetary.org/news/2006/0116_New_Horizons_Set_to_Launch_on_9_Year.html). *The Planetary Society*. . Retrieved 2006-11-08.

[9] "Dawn: A Journey to the Beginning of the Solar System" (http://www-ssc.igpp.ucla.edu/dawn/background.html). *Space Physics Center: UCLA*. 2005. . Retrieved 2007-11-03.

[10] "The Sun's Vital Statistics" (http://solar-center.stanford.edu/vitalstats.html). Stanford Solar Center. . Retrieved 2008-07-29., citing Eddy, J. (1979). *A New Sun: The Solar Results From Skylab* (http://history.nasa.gov/SP-402/contents.htm). NASA. p. 37. NASA SP-402. .

[11] Williams, Dr. David R. (September 7, 2006). "Saturn Fact Sheet" (http://nssdc.gsfc.nasa.gov/planetary/factsheet/saturnfact.html). NASA. . Retrieved 2007-07-31.

[12] Williams, Dr. David R. (November 16, 2004). "Jupiter Fact Sheet" (http://nssdc.gsfc.nasa.gov/planetary/factsheet/jupiterfact.html). NASA. . Retrieved 2007-08-08.

[13] Paul Robert Weissman, Torrence V. Johnson (2007). *Encyclopedia of the solar system*. Academic Press. p. 615. ISBN 0120885891.

[14] "Planet Formation (in the Solar System)" (http://www.astro.utoronto.ca/~mhvk/AST221/L20/L20_4.pdf). University of Toronto. . Retrieved 2011-07-11.

[15] Podolak, M.; Weizman, A.; Marley, M. (1995). "Comparative models of Uranus and Neptune". *Planetary and Space Sciences* **43** (12): 1517–1522. Bibcode 1995P&SS...43.1517P. doi:10.1016/0032-0633(95)00061-5.

[16] Podolak, M.; Podolak, J.I.; Marley, M.S. (2000). "Further investigations of random models of Uranus and Neptune". *Planetary & Spaces Sciences* **48** (2–3): 143–151. Bibcode 2000P&SS...48..143P. doi:10.1016/S0032-0633(99)00088-4.

[17] Michael Zellik (2002). *Astronomy: The Evolving Universe* (9th ed.). Cambridge University Press. p. 240. ISBN 0521800900. OCLC 223304585 46685453.

[18] Placxo, Kevin W.; Gross, Michael (2006). *Astrobiology: a brief introduction* (http://books.google.com/?id=2JuGDL144BEC&pg=PA66&dq=inventory+volatiles+hydrogen&q=inventory volatiles hydrogen). JHU Press. p. 66. ISBN 9780801883675. .

[19] "Sun: Facts & Figures" (http://web.archive.org/web/20080102034758/http://solarsystem.nasa.gov/planets/profile.cfm?Object=Sun&Display=Facts&System=Metric). NASA. Archived from the original (http://solarsystem.nasa.gov/planets/profile.cfm?Object=Sun&Display=Facts&System=Metric) on 2008-01-02. . Retrieved 2009-05-14.

[20] Zirker, Jack B. (2002). *Journey from the Center of the Sun*. Princeton University Press. pp. 120–127. ISBN 9780691057811.

[21] "Why is visible light visible, but not other parts of the spectrum?" (http://www.straightdope.com/columns/read/2085/why-is-visible-light-visible-but-not-other-parts-of-the-spectrum). The Straight Dome. 2003. . Retrieved 2009-05-14.

[22] Than, Ker (January 30, 2006). "Astronomers Had it Wrong: Most Stars are Single" (http://www.space.com/scienceastronomy/060130_mm_single_stars.html). SPACE.com. . Retrieved 2007-08-01.

[23] Smart, R. L.; Carollo, D.; Lattanzi, M. G.; McLean, B.; Spagna, A. (2001). "The Second Guide Star Catalogue and Cool Stars". In Hugh R. A. Jones and Iain A. Steele. *Ultracool Dwarfs: New Spectral Types L and T*. Springer. pp. 119. Bibcode 2001udns.conf..119S.

[24] Nir J. Shaviv (2003). "Towards a Solution to the Early Faint Sun Paradox: A Lower Cosmic Ray Flux from a Stronger Solar Wind". *Journal of Geophysical Research* **108** (A12): 1437. arXiv:astroph/0306477. Bibcode 2003JGRA..108.1437S. doi:10.1029/2003JA009997.

[25] T. S. van Albada, Norman Baker (1973). "On the Two Oosterhoff Groups of Globular Clusters". *Astrophysical Journal* **185**: 477–498. Bibcode 1973ApJ...185..477V. doi:10.1086/152434.

[26] Charles H. Lineweaver (2001-03-09). "An Estimate of the Age Distribution of Terrestrial Planets in the Universe: Quantifying Metallicity as a Selection Effect". *Icarus* **151** (2): 307–313. Bibcode 2001Icar..151..307L. doi:10.1006/icar.2001.6607.

[27] "Solar Physics: The Solar Wind" (http://solarscience.msfc.nasa.gov/SolarWind.shtml). *Marshall Space Flight Center*. 2006-07-16. . Retrieved 2006-10-03.

[28] "Voyager Enters Solar System's Final Frontier" (http://www.nasa.gov/vision/universe/solarsystem/voyager_agu.html). *NASA*. . Retrieved 2007-04-02.

[29] Phillips, Tony (2001-02-15). "The Sun Does a Flip" (http://science.nasa.gov/headlines/y2001/ast15feb_1.htm). *Science@NASA*. . Retrieved 2007-02-04.

[30] A Star with two North Poles (http://science.nasa.gov/headlines/y2003/22apr_currentsheet.htm), April 22, 2003, Science @ NASA

[31] Riley, Pete (2002). "Modeling the heliospheric current sheet: Solar cycle variations" (http://ulysses.jpl.nasa.gov/science/monthly_highlights/2002-July-2001JA000299.pdf). *Journal of Geophysical Research* **107**. Bibcode 2002JGRA.107g.SSH8R. doi:10.1029/2001JA000299. .

[32] Lundin, Richard (2001-03-09). "Erosion by the Solar Wind". *Science* **291** (5510): 1909. doi:10.1126/science.1059763. PMID 11245195.

[33] Langner, U. W.; M. S. Potgieter (2005). "Effects of the position of the solar wind termination shock and the heliopause on the heliospheric modulation of cosmic rays". *Advances in Space Research* **35** (12): 2084–2090. Bibcode 2005AdSpR..35.2084L. doi:10.1016/j.asr.2004.12.005.

[34] "Long-term Evolution of the Zodiacal Cloud" (http://astrobiology.arc.nasa.gov/workshops/1997/zodiac/backman/IIIc.html). 1998. . Retrieved 2007-02-03.

[35] "ESA scientist discovers a way to shortlist stars that might have planets" (http://sci.esa.int/science-e/www/object/index.cfm?fobjectid=29471). *ESA Science and Technology*. 2003. . Retrieved 2007-02-03.

[36] Landgraf, M.; Liou, J.-C.; Zook, H. A.; Grün, E. (May 2002). "Origins of Solar System Dust beyond Jupiter" (http://astron.berkeley.edu/~kalas/disksite/library/ladgraf02.pdf). *The Astronomical Journal* **123** (5): 2857–2861. Bibcode 2002AJ....123.2857L. doi:10.1086/339704. . Retrieved 2007-02-09.

[37] "Inner Solar System" (http://nasascience.nasa.gov/planetary-science/exploring-the-inner-solar-system). NASA Science (Planets). . Retrieved 2009-05-09.

[38] Schenk P., Melosh H. J. (1994), *Lobate Thrust Scarps and the Thickness of Mercury's Lithosphere*, Abstracts of the 25th Lunar and Planetary Science Conference, 1994LPI....25.1203S

[39] Bill Arnett (2006). "Mercury" (http://www.nineplanets.org/mercury.html). *The Nine Planets*. . Retrieved 2006-09-14.

[40] Benz, W., Slattery, W. L., Cameron, A. G. W. (1988), *Collisional stripping of Mercury's mantle*, Icarus, v. 74, p. 516–528.

[41] Cameron, A. G. W. (1985), *The partial volatilization of Mercury*, Icarus, v. 64, p. 285–294.

[42] Mark Alan Bullock (1997) (PDF). *The Stability of Climate on Venus* (http://www.boulder.swri.edu/~bullock/Homedocs/PhDThesis.pdf). Southwest Research Institute. . Retrieved 2006-12-26.

[43] Paul Rincon (1999). "Climate Change as a Regulator of Tectonics on Venus" (http://www.boulder.swri.edu/~bullock/Homedocs/Science2_1999.pdf) (PDF). *Johnson Space Center Houston, TX, Institute of Meteoritics, University of New Mexico, Albuquerque, NM*. . Retrieved 2006-11-19.

[44] "What are the characteristics of the Solar System that lead to the origins of life?" (http://science.nasa.gov/planetary-science/big-questions/what-are-the-characteristics-of-the-solar-system-that-lead-to-the-origins-of-life-1/). NASA Science (Big Questions). . Retrieved 2011-08-30.

[45] Anne E. Egger, M.A./M.S.. "Earth's Atmosphere: Composition and Structure" (http://www.visionlearning.com/library/module_viewer.php?c3=&mid=107&l=). *VisionLearning.com*. . Retrieved 2006-12-26.

[46] David C. Gatling, Conway Leovy (2007). "Mars Atmosphere: History and Surface Interactions". In Lucy-Ann McFadden et. al.. *Encyclopaedia of the Solar System*. pp. 301–314.

[47] David Noever (2004). "Modern Martian Marvels: Volcanoes?" (http://www.astrobio.net/news/modules.php?op=modload&name=News&file=article&sid=1360&mode=thread&order=0&thold=0). *NASA Astrobiology Magazine*. . Retrieved 2006-07-23.

[48] "Mars: A Kid's Eye View" (http://solarsystem.nasa.gov/planets/profile.cfm?Object=Mars&Display=Kids). NASA. . Retrieved 2009-05-14.

[49] Scott S. Sheppard, David Jewitt, and Jan Kleyna (2004). "A Survey for Outer Satellites of Mars: Limits to Completeness" (http://www2.ess.ucla.edu/~jewitt/papers/2004/SJK2004.pdf). *Astronomical Journal*. . Retrieved 2006-12-26.

[50] "Are Kuiper Belt Objects asteroids? Are large Kuiper Belt Objects planets?" (http://curious.astro.cornell.edu/question.php?number=601). Cornell University. . Retrieved 2009-03-01.

[51] Petit, J.-M.; Morbidelli, A.; Chambers, J. (2001). "The Primordial Excitation and Clearing of the Asteroid Belt" (http://www.gps.caltech.edu/classes/ge133/reading/asteroids.pdf) (PDF). *Icarus* **153** (2): 338–347. Bibcode 2001Icar..153..338P. doi:10.1006/icar.2001.6702. . Retrieved 2007-03-22.

[52] "IAU Planet Definition Committee" (http://www.iau.org/public_press/news/release/iau0601/newspaper/). International Astronomical Union. 2006. . Retrieved 2009-03-01.

[53] "New study reveals twice as many asteroids as previously believed" (http://www.esa.int/esaCP/ESAASPF18ZC_index_0.html). *ESA*. 2002. . Retrieved 2006-06-23.

[54] Krasinsky, G. A.; Pitjeva, E. V.; Vasilyev, M. V.; Yagudina, E. I. (July 2002). "Hidden Mass in the Asteroid Belt". *Icarus* **158** (1): 98–105. Bibcode 2002Icar..158...98K. doi:10.1006/icar.2002.6837.

[55] Beech, M.; Steel; Duncan I. Steel (September 1995). "On the Definition of the Term Meteoroid". *Quarterly Journal of the Royal Astronomical Society* **36** (3): 281–284. Bibcode 1995QJRAS..36..281B.

[56] "History and Discovery of Asteroids" (http://dawn.jpl.nasa.gov/DawnClassrooms/1_hist_dawn/history_discovery/Development/a_modeling_scale.doc) (DOC). *NASA*. . Retrieved 2006-08-29.

[57] Phil Berardelli (2006). "Main-Belt Comets May Have Been Source Of Earths Water" (http://www.spacedaily.com/reports/Main_Belt_Comets_May_Have_Been_Source_Of_Earths_Water.html). *SpaceDaily*. . Retrieved 2006-06-23.

[58] Barucci, M. A.; Kruikshank, D.P.; Mottola S.; Lazzarin M. (2002). "Physical Properties of Trojan and Centaur Asteroids". *Asteroids III*. Tucson, Arizona: University of Arizona Press. pp. 273–87.

[59] A. Morbidelli, W. F. Bottke Jr., Ch. Froeschlé, P. Michel; Bottke; Froeschlé; Michel (January 2002). W. F. Bottke Jr., A. Cellino, P. Paolicchi, and R. P. Binzel. ed. "Origin and Evolution of Near-Earth Objects" (http://www.boulder.swri.edu/~bottke/Reprints/Morbidelli-etal_2002_AstIII_NEOs.pdf) (PDF). *Asteroids III* (University of Arizona Press): 409–422. Bibcode 2002aste.conf..409M. .

[60] Jack J. Lissauer, David J. Stevenson (2006). "Formation of Giant Planets" (http://www.gps.caltech.edu/uploads/File/People/djs/lissauer&stevenson(PPV).pdf) (PDF). *NASA Ames Research Center; California Institute of Technology*. . Retrieved 2006-01-16.

[61] Pappalardo, R T (1999). "Geology of the Icy Galilean Satellites: A Framework for Compositional Studies" (http://www.agu.org/cgi-bin/SFgate/SFgate?&listenv=table&multiple=1&range=1&directget=1&application=fm99&database=/data/epubs/wais/indexes/fm99/fm99&maxhits=200&="P11C-10"). *Brown University*. . Retrieved 2006-01-16.

[62] Kargel, J. S. (1994). "Cryovolcanism on the icy satellites". *Earth, Moon, and Planets* **67**: 101–113. Bibcode 1995EM&P...67..101K. doi:10.1007/BF00613296.

[63] Hawksett, David; Longstaff, Alan; Cooper, Keith; Clark, Stuart; Longstaff; Cooper; Clark (2005). "10 Mysteries of the Solar System". *Astronomy Now* **19**: 65. Bibcode 2005AsNow..19h..65H.

[64] Podolak, M.; Reynolds, R. T.; Young, R. (1990). "Post Voyager comparisons of the interiors of Uranus and Neptune". *Geophysical Research Letters* **17** (10): 1737. Bibcode 1990GeoRL..17.1737P. doi:10.1029/GL017i010p01737.

[65] Duxbury, N. S., Brown, R. H. (1995). "The Plausibility of Boiling Geysers on Triton" (http://trs-new.jpl.nasa.gov/dspace/handle/2014/28034?mode=full). *Beacon eSpace*. . Retrieved 2006-01-16.

[66] Sekanina, Zdeněk (2001). "Kreutz sungrazers: the ultimate case of cometary fragmentation and disintegration?". *Publications of the Astronomical Institute of the Academy of Sciences of the Czech Republic* **89**: 78–93. Bibcode 2001PAICz..89...78S.

[67] Królikowska, M. (2001). "A study of the original orbits of *hyperbolic* comets". *Astronomy & Astrophysics* **376** (1): 316–324. Bibcode 2001A&A...376..316K. doi:10.1051/0004-6361:20010945.

[68] Whipple, Fred L. (1992). "The activities of comets related to their aging and origin". *Celestial Mechanics and Dynamical Astronomy* **54**: 1–11. Bibcode 1992CeMDA..54....1W. doi:10.1007/BF00049540.

[69] John Stansberry, Will Grundy, Mike Brown, Dale Cruikshank, John Spencer, David Trilling, Jean-Luc Margot (2007). "Physical Properties of Kuiper Belt and Centaur Objects: Constraints from Spitzer Space Telescope". *The Solar System Beyond Neptune*. pp. 161. arXiv:astro-ph/0702538. Bibcode 2008ssbn.book..161S.

[70] Patrick Vanouplines (1995). "Chiron biography" (http://www.vub.ac.be/STER/www.astro/chibio.htm). *Vrije Universitiet Brussel*. . Retrieved 2006-06-23.

[71] Stephen C. Tegler (2007). "Kuiper Belt Objects: Physical Studies". In Lucy-Ann McFadden et. al.. *Encyclopedia of the Solar System*. pp. 605–620.

[72] Audrey Delsanti and David Jewitt (2006). "The Solar System Beyond The Planets" (http://web.archive.org/web/20070129151907/http://www.ifa.hawaii.edu/faculty/jewitt/papers/2006/DJ06.pdf) (PDF). *Institute for Astronomy, University of Hawaii*. Archived from the original (http://www.ifa.hawaii.edu/faculty/jewitt/papers/2006/DJ06.pdf) on January 29, 2007. . Retrieved 2007-01-03.

[73]
This citation will be automatically completed in the next few minutes. You can jump the queue or expand by hand (http://en.wikipedia.org/wiki/Template:cite_doi/_10.1086.2f501524_)

[74] Chiang *et al.*; Jordan, A. B.; Millis, R. L.; Buie, M. W.; Wasserman, L. H.; Elliot, J. L.; Kern, S. D.; Trilling, D. E. et al. (2003). "Resonance Occupation in the Kuiper Belt: Case Examples of the 5:2 and Trojan Resonances" (http://www.boulder.swri.edu/~buie/biblio/pub047.pdf). *The Astronomical Journal* **126** (1): 430–443. Bibcode 2003AJ....126..430C. doi:10.1086/375207. . Retrieved 2009-08-15.

[75] M. W. Buie, R. L. Millis, L. H. Wasserman, J. L. Elliot, S. D. Kern, K. B. Clancy, E. I. Chiang, A. B. Jordan, K. J. Meech, R. M. Wagner, D. E. Trilling (2005). "Procedures, Resources and Selected Results of the Deep Ecliptic Survey". *Earth, Moon, and Planets* **92** (1): 113.

arXiv:astro-ph/0309251. Bibcode 2003EM&P...92..113B. doi:10.1023/B:MOON.0000031930.13823.be.

[76] E. Dottol, M. A. Barucci2, and M. Fulchignoni (2006-08-24). "Beyond Neptune, the new frontier of the Solar System" (http://sait.oat.ts.astro.it/MSAIS/3/PDF/20.pdf) (PDF). . Retrieved 2006-12-26.

[77] Fajans, J.; L. Frièdland (October 2001). "Autoresonant (nonstationary) excitation of pendulums, Plutinos, plasmas, and other nonlinear oscillators" (http://ist-socrates.berkeley.edu/~fajans/pub/pdffiles/AutoPendAJP.pdf). *American Journal of Physics* **69** (10): 1096–1102. doi:10.1119/1.1389278. . Retrieved 2006-12-26.

[78] "Dwarf Planets and their Systems" (http://planetarynames.wr.usgs.gov/append7.html#DwarfPlanets). *Working Group for Planetary System Nomenclature (WGPSN)*. U.S. Geological Survey. 2008-11-07. . Retrieved 2008-07-13.

[79] Marc W. Buie (2008-04-05). "Orbit Fit and Astrometric record for 136472" (http://www.boulder.swri.edu/~buie/kbo/astrom/136472.html). SwRI (Space Science Department). . Retrieved 2008-07-13.

[80] David Jewitt (2005). "The 1000 km Scale KBOs" (http://www2.ess.ucla.edu/~jewitt/kb/big_kbo.html). *University of Hawaii*. . Retrieved 2006-07-16.

[81] "List Of Centaurs and Scattered-Disk Objects" (http://www.minorplanetcenter.org/iau/lists/Centaurs.html). *IAU: Minor Planet Center*. . Retrieved 2007-04-02.

[82] Michael E. Brown and Emily L. Schaller (2007). "The Mass of Dwarf Planet Eris". *Science* **316** (5831): 1585. Bibcode 2007Sci...316.1585B. doi:10.1126/science.1139415. PMID 17569855.

[83] Littmann, Mark (2004). *Planets Beyond: Discovering the Outer Solar System*. Courier Dover Publications. pp. 162–163. ISBN 9780486436029.

[84] Fahr, H. J.; Kausch, T.; Scherer, H.; Kausch; Scherer (2000). "A 5-fluid hydrodynamic approach to model the Solar System-interstellar medium interaction" (http://aa.springer.de/papers/0357001/2300268.pdf) (PDF). *Astronomy & Astrophysics* **357**: 268. Bibcode 2000A&A...357..268F. . See Figures 1 and 2.

[85] NASA/JPL (2009). "Cassini's Big Sky: The View from the Center of Our Solar System" (http://www.jpl.nasa.gov/news/features.cfm?feature=2370&msource=F20091119&tr=y&auid=5615216). . Retrieved 2009-12-20.

[86] Stone, E. C.; Cummings, A. C.; McDonald, F. B.; Heikkila, B. C.; Lal, N.; Webber, W. R. (September 2005). "Voyager 1 explores the termination shock region and the heliosheath beyond". *Science* **309** (5743): 2017–20. Bibcode 2005Sci...309.2017S. doi:10.1126/science.1117684. PMID 16179468.

[87] Stone, E. C.; Cummings, A. C.; McDonald, F. B.; Heikkila, B. C.; Lal, N.; Webber, W. R. (July 2008). "An asymmetric solar wind termination shock". *Nature* **454** (7200): 71–4. doi:10.1038/nature07022. PMID 18596802.

[88] P. C. Frisch (University of Chicago) (June 24, 2002). "The Sun's Heliosphere & Heliopause" (http://antwrp.gsfc.nasa.gov/apod/ap020624.html). *Astronomy Picture of the Day*. . Retrieved 2006-06-23.

[89] "Voyager: Interstellar Mission" (http://voyager.jpl.nasa.gov/mission/interstellar.html). *NASA Jet Propulsion Laboratory*. 2007. . Retrieved 2008-05-08.

[90] R. L. McNutt, Jr. et al. (2006). "Innovative Interstellar Explorer". *Physics of the Inner Heliosheath: Voyager Observations, Theory, and Future Prospects*. AIP Conference Proceedings. **858**. pp. 341–347. Bibcode 2006AIPC..858..341M. doi:10.1063/1.2359348.

[91] Anderson, Mark (2007-01-05). "Interstellar space, and step on it!" (http://space.newscientist.com/article/mg19325850.900-interstellar-space-and-step-on-it.html). *New Scientist*. . Retrieved 2007-02-05.

[92] Stern SA, Weissman PR. (2001). "Rapid collisional evolution of comets during the formation of the Oort cloud." (http://www.ncbi.nlm.nih.gov/entrez/query.fcgi?cmd=Retrieve&db=PubMed&list_uids=11214311&dopt=Citation). *Space Studies Department, Southwest Research Institute, Boulder, Colorado*. . Retrieved 2006-11-19.

[93] Bill Arnett (2006). "The Kuiper Belt and the Oort Cloud" (http://www.nineplanets.org/kboc.html). *nineplanets.org*. . Retrieved 2006-06-23.

[94] David Jewitt (2004). "Sedna – 2003 VB_{12}" (http://www2.ess.ucla.edu/~jewitt/kb/sedna.html). *University of Hawaii*. . Retrieved 2006-06-23.

[95] Mike Brown. "Sedna" (http://www.gps.caltech.edu/~mbrown/sedna/). *CalTech*. . Retrieved 2007-05-02.

[96] T. Encrenaz, JP. Bibring, M. Blanc, MA. Barucci, F. Roques, PH. Zarka (2004). *The Solar System: Third edition*. Springer. p. 1.

[97] Durda D. D.; Stern S. A.; Colwell W. B.; Parker J. W.; Levison H. F.; Hassler D. M. (2004). "A New Observational Search for Vulcanoids in SOHO/LASCO Coronagraph Images". *Icarus* **148**: 312–315. Bibcode 2000Icar..148..312D. doi:10.1006/icar.2000.6520.

[98] English, J. (2000). "Exposing the Stuff Between the Stars" (http://www.ras.ucalgary.ca/CGPS/press/aas00/pr/pr_14012000/pr_14012000map1.html) (Press release). Hubble News Desk. . Retrieved 2007-05-10.

[99] R. Drimmel, D. N. Spergel (2001). "Three Dimensional Structure of the Milky Way Disk". *Astrophysical Journal* **556**: 181–202. arXiv:astro-ph/0101259. Bibcode 2001ApJ...556..181D. doi:10.1086/321556.

[100] Eisenhauer, F.; et al. (2003). "A Geometric Determination of the Distance to the Galactic Center". *Astrophysical Journal* **597** (2): L121–L124. Bibcode 2003ApJ...597L.121E. doi:10.1086/380188.

[101] Leong, Stacy (2002). "Period of the Sun's Orbit around the Galaxy (Cosmic Year)" (http://hypertextbook.com/facts/2002/StacyLeong.shtml). *The Physics Factbook*. . Retrieved 2007-04-02.

[102] C. Barbieri (2003). "Elementi di Astronomia e Astrofisica per il Corso di Ingegneria Aerospaziale V settimana" (http://dipastro.pd.astro.it/planets/barbieri/Lezioni-AstroAstrofIng04_05-Prima-Settimana.ppt). *IdealStars.com*. . Retrieved 2007-02-12.

[103] Leslie Mullen (2001). "Galactic Habitable Zones" (http://www.astrobio.net/news/modules.php?op=modload&name=News&file=article&sid=139). *Astrobiology Magazine*. . Retrieved 2006-06-23.

[104] "Supernova Explosion May Have Caused Mammoth Extinction" (http://www.physorg.com/news6734.html). *Physorg.com*. 2005. . Retrieved 2007-02-02.

[105] "Near-Earth Supernovas" (http://science.nasa.gov/headlines/y2003/06jan_bubble.htm). *NASA*. . Retrieved 2006-07-23.

[106] "Stars within 10 light years" (http://www.solstation.com/stars/s10ly.htm). *SolStation*. . Retrieved 2007-04-02.

[107] "Tau Ceti" (http://www.solstation.com/stars/tau-ceti.htm). *SolStation*. . Retrieved 2007-04-02.

[108] "HUBBLE ZEROES IN ON NEAREST KNOWN EXOPLANET" (http://hubblesite.org/newscenter/archive/releases/2006/32/text/). *Hubblesite*. 2006. . Retrieved 2008-01-13.

[109] http://upload.wikimedia.org/wikipedia/commons/0/0f/Earth%27s_Location_in_the_Universe_SMALLER_%28JPEG%29.jpg

[110] The date is based on the oldest inclusions found to date in meteorites, and is thought to be the date of the formation of the first solid material in the collapsing nebula.

A. Bouvier and M. Wadhwa. "The age of the solar system redefined by the oldest Pb-Pb age of a meteoritic inclusion." *Nature Geoscience,* in press, 2010. Doi: 10.1038/NGEO941

[111] "Lecture 13: The Nebular Theory of the origin of the Solar System" (http://atropos.as.arizona.edu/aiz/teaching/nats102/mario/solar_system.html). *University of Arizona*. . Retrieved 2006-12-27.

[112] Irvine, W. M. (1983). "The chemical composition of the pre-solar nebula". *Cometary exploration; Proceedings of the International Conference*. **1**. pp. 3. Bibcode 1983coex....1....3I.

[113] Greaves, Jane S. (2005-01-07). "Disks Around Stars and the Growth of Planetary Systems". *Science* **307** (5706): 68–71. Bibcode 2005Sci...307...68G. doi:10.1126/science.1101979. PMID 15637266.

[114] "Present Understanding of the Origin of Planetary Systems" (http://www7.nationalacademies.org/ssb/detectionch3.html). National Academy of Sciences. 2000-04-05. . Retrieved 2007-01-19.

[115] M. Momose, Y. Kitamura, S. Yokogawa, R. Kawabe, M. Tamura, S. Ida (2003). "Investigation of the Physical Properties of Protoplanetary Disks around T Tauri Stars by a High-resolution Imaging Survey at lambda = 2 mm". In Ikeuchi, S., Hearnshaw, J. and Hanawa, T. (eds.). *The Proceedings of the IAU 8th Asian-Pacific Regional Meeting, Volume I*. ASP Conference Series. **289**. pp. 85. Bibcode 2003ASPC..289...85M.

[116] Boss, A. P.; Durisen, R. H. (2005). "Chondrule-forming Shock Fronts in the Solar Nebula: A Possible Unified Scenario for Planet and Chondrite Formation". *The Astrophysical Journal* **621** (2): L137. Bibcode 2005ApJ...621L.137B. doi:10.1086/429160.

[117] Sukyoung Yi; Pierre Demarque; Yong-Cheol Kim; Young-Wook Lee; Chang H. Ree; Thibault Lejeune; Sydney Barnes (2001). "Toward Better Age Estimates for Stellar Populations: The Isochrones for Solar Mixture". *Astrophysical Journal Supplement* **136**: 417. arXiv:astro-ph/0104292. Bibcode 2001ApJS..136..417Y. doi:10.1086/321795.

[118] A. Chrysostomou, P. W. Lucas (2005). "The Formation of Stars". *Contemporary Physics* **46** (1): 29. Bibcode 2005ConPh..46...29C. doi:10.1080/0010751042000275277.

[119] Jeff Hecht (1994). "Science: Fiery future for planet Earth" (http://www.newscientist.com/article/mg14219191.900.html). *NewScientist*. . Retrieved 2007-10-29.

[120] K. P. Schroder, Robert Cannon Smith (2008). "Distant future of the Sun and Earth revisited". *Monthly Notices of the Royal Astronomical Society* **386** (1): 155–163. Bibcode 2008MNRAS.386..155S. doi:10.1111/j.1365-2966.2008.13022.x.

[121] Pogge, Richard W. (1997). "The Once & Future Sun" (http://web.archive.org/web/20050527094435/http://www-astronomy.mps.ohio-state.edu/Vistas/) (lecture notes). *New Vistas in Astronomy* (http://www-astronomy.mps.ohio-state.edu/Vistas/). Archived from the original (http://www.astronomy.ohio-state.edu/~pogge/Lectures/vistas97.html) on May 27, 2005. . Retrieved 2005-12-07.

[122] http://www.iau.org/public_press/themes/naming/

[123] http://www.m-w.com/dictionary/solar%20system

[124] Alessandro Morbidelli (2005). "Origin and dynamical evolution of comets and their reservoirs". arXiv:astro-ph/0512256.

[125] "The Final IAU Resolution on the definition of "planet" ready for voting" (http://www.iau.org/iau0602.423.0.html). IAU. 2006-08-24. . Retrieved 2007-03-02.

[126] Ron Ekers. "IAU Planet Definition Committee" (http://www.iau.org/public_press/news/release/iau0601/newspaper/). International Astronomical Union. . Retrieved 2008-10-13.

[127] "Plutoid chosen as name for Solar System objects like Pluto" (http://www.iau.org/public_press/news/release/iau0804). International Astronomical Union. June 11, 2008, Paris. . Retrieved 2008-06-11.

[128] Reid, M.J.; Brunthaler, A. (2004 2004). "The Proper Motion of Sagittarius A*". *The Astrophysical Journal* **616** (2): 883. Bibcode 2004ApJ...616..872R. doi:10.1086/424960.

External links

- Solar System Profile (http://solarsystem.nasa.gov/planets/profile.cfm?Object=SolarSys&Display=Overview) by NASA's Solar System Exploration (http://solarsystem.nasa.gov/index.cfm)
- NASA's Solar System Simulator (http://space.jpl.nasa.gov)
- NASA/JPL Solar System main page (http://www.jpl.nasa.gov/solar_system)

bjn:Tata Surya rue:Сонячна сістема

Near-Earth_object

A **near-Earth object** (**NEO**) is a Solar System object whose orbit brings it into close proximity with the Earth. All NEOs have a perihelion distance less than 1.3 AU.[1] They include a few thousand near-Earth asteroids (NEAs), near-Earth comets, a number of solar-orbiting spacecraft, and meteoroids large enough to be tracked in space before striking the Earth. It is now widely accepted that collisions in the past have had a significant role in shaping the geological and biological history of the planet.[2] NEOs have become of increased interest since the 1980s because of increased awareness of the potential danger some of the asteroids or comets pose to the Earth, and active mitigations are being researched. A study showed that the United States and China are the nations most vulnerable to a meteor strike.[3]

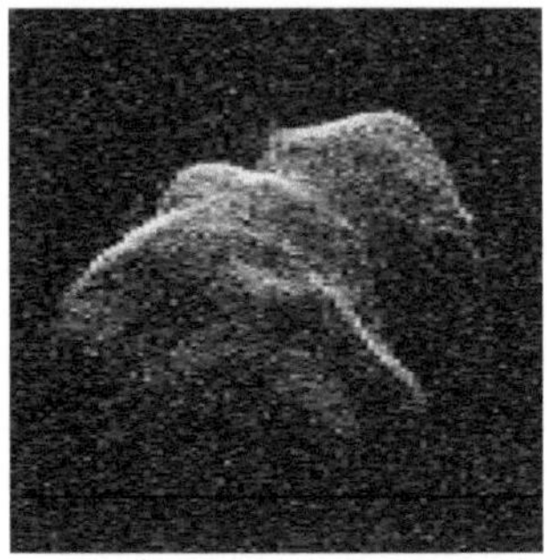

Asteroid 4179 Toutatis is a potentially hazardous object that has passed within 2.3 lunar distances.

Those NEOs that are asteroids (NEA) have orbits that lie partly between 0.983 and 1.3 astronomical units away from the Sun.[4] When an NEA is detected it is submitted to the Harvard Minor Planet Center for cataloging. Some near-Earth asteroids' orbits intersect that of Earth's so they pose a collision danger.[5] The United States, European Union and other nations are currently scanning for NEOs[6] in an effort called Spaceguard.

Asteroid Toutatis from Paranal.

In the United States, NASA has a congressional mandate to catalogue all NEOs that are at least 1 kilometer wide, as the impact of such an object would be expected to produce severe to catastrophic effects. As of October 2008, 982 of these mandated NEOs have been detected.[7] It was estimated in 2006 that 20% of the mandated objects have not yet been found.[6] Efforts are under way to use an existing telescope in Australia to cover the ~30% of the sky that has not yet been surveyed.

Potentially hazardous objects (PHOs) are currently defined based on parameters that measure the object's potential to make threatening close approaches to the Earth.[8] Mostly objects with an Earth minimum orbit intersection distance (MOID) of 0.05 AU or less and an absolute magnitude (H) of 22.0 or less (a rough indicator of large size) are considered PHOs. Objects that cannot approach closer to the Earth (i.e. MOID) than 0.05 AU (km; mi), or are smaller than about 150 m (500 ft) in diameter (i.e. H = 22.0 with assumed albedo of 13%), are not considered PHOs.[9] The NASA Near Earth Object Catalog also includes the approach distances of asteroids and comets measured in Lunar Distances,[10] and this usage has become the more usual unit of measure used by the press and mainstream media in discussing these objects.

Some NEOs are of high interest because they can be physically explored with lower mission velocity even than the Moon, due to their combination of low velocity with respect to Earth (ΔV) and small gravity, so they may present interesting scientific opportunities both for direct geochemical and astronomical investigation, and as potentially economical sources of extraterrestrial materials for human exploitation.[11] This makes them an attractive target for exploration.[12] As of 2008, two near-Earth objects have been visited by spacecraft: 433 Eros, by NASA's Near Earth Asteroid Rendezvous probe,[13] and 25143 Itokawa, by the JAXA Hayabusa mission.[14]

Risk scales

There are two schemes for classification of impact hazards:

- the simple Torino Scale, and
- the more complex Palermo Technical Impact Hazard Scale

The annual background frequency used in the Palermo scale for impacts of energy greater than *E* megatonnes is estimated as:

For instance, this formula implies that the expected value of the time from now until the next impact greater than 1 megatonne is 33 years, and that when it occurs, there is a 50% chance that it will be above 2.4 megatonnes. This formula is only valid over a certain range of *E*.

However, another paper[15] published in 2002 – the same year as the paper on which the Palermo scale is based – found a power law with different constants:

This formula gives considerably lower rates for a given *E*. For instance, it gives the rate for bolides of 10 megatonnes or more (like the Tunguska explosion) as 1 per thousand years, rather than 1 per 210 years as in the Palermo formula. However, the authors give a rather large uncertainty (once in 400 to 1800 years for 10 megatonnes), due in part to uncertainties in determining the energies of the atmospheric impacts that they used in their determination.

Highly rated risks

On 25 December 2004, minor planet 2004 MN_4 (later named 99942 Apophis) was assigned a 4 on the Torino scale, the highest rating so far. On 27 December 2004, there was a 2.7% chance of Earth impact on 13 April 2029. However, on 28 December 2004, the risk of impact dropped to zero for 2029, but, due to a resonant return possibility the Torino rating for an April 2036 impact rose to 4 in early 2005, but by October 2009 the Torino rating was 0 (zero). The Palermo rating (October 2009) is –3.08.[16]

As of 12 September 2011, the only known NEO with a Palermo scale value greater than zero is (29075) 1950 DA, which is predicted to pass very close to or collide with the Earth ($p \leq 0.003$) in the year 2880. Depending on the orientation of its axis of rotation, it will either miss the Earth by tens of millions of kilometers, or have a 1 in 300 chance of hitting the Earth. However, humanity has over 800 years to refine its estimates of the orbit of (29075) 1950 DA, and to deflect it, if necessary.[17]

List of current threats

NASA maintains a continuously updated web page of the most significant NEO threats in the next 100 years.[16] All or nearly all of the items on this page are highly likely to drop off the list eventually as more data comes in, enabling more accurate predictions. (The page does not include 1950 DA, because that will not strike for at least 800 years.)

Number and classification of near-Earth objects

While orbiting the Sun, most potential impactors can be classified as meteoroids, asteroids, or comets depending on size and composition. Asteroids can also be members of an asteroid family, and comets can leave debris in their orbits.

As of April 2011, 7,954 NEOs have been discovered: 87 near-Earth comets and 7,867 near-Earth Asteroids. Of those there are 647 Aten asteroids, 2,920 Amor asteroids, and 4,289 Apollo asteroids. There are 1,215 NEOs that are classified as potentially hazardous asteroids (PHAs). Currently, 148 PHAs and 824 NEAs have an absolute magnitude of 17.75 or brighter, which roughly corresponds to at least 1 km in size.[18]

As of April 2011, there are 368 NEAs on the impact risk page at the NASA website.[16] A significant number of these NEAs – 215 as of May 2010 – are equal to or smaller than 50 meters in diameter and none of the listed objects are placed even in the "yellow zone" (Torino Scale 2), meaning that none warrant the attention of general public.[19] As of November 2011, only asteroids 2007 VK_{184} and 2011 AG_5 are listed as having a Torino Scale of 1.

Near-Earth meteoroids

Near-Earth meteoroids are objects with orbits in the vicinity of Earth's orbit having a diameter less than 50 meters.

Near-Earth asteroids

These are objects of 50 meters or more in diameter in a near-Earth orbit without the tail or coma of a comet. As of May 2010, 7,075 near-Earth asteroids are known,[18] ranging in size up to ~32 kilometers (1036 Ganymed).[20] The number of near-Earth asteroids over one kilometer in diameter is estimated to be 500–1,000.[21] [22] The composition of near-Earth asteroids is comparable to that of asteroids from the asteroid belt, reflecting a variety of asteroid spectral types.[23]

There are significantly fewer near-Earth asteroids in the mid-size range than previously thought.

NEAs survive in their orbits for just a few million years.[4] They are eventually eliminated by planetary perturbations which cause ejection from the Solar System or a collision with the Sun or a planet. With orbital lifetimes short compared to the age of the Solar System, new asteroids must be constantly moved into near-Earth orbits to explain the observed asteroids. The accepted origin of these asteroids is that asteroid-belt asteroids are moved into the inner Solar System through orbital resonances with Jupiter. The interaction with Jupiter through the resonance perturbs the asteroid's orbit and it comes into the inner Solar System. The asteroid belt has gaps, known as Kirkwood gaps, where these resonances occur as the asteroids in these resonances have been moved onto other orbits. New asteroids migrate into these resonances, due to the Yarkovsky effect that provides a continuing supply of near-Earth asteroids.[24]

A small number of NEOs are extinct comets that have lost their volatile surface materials, although having a faint or intermittent comet-like tail does not necessarily result in a classification as a near-Earth comet, making the boundaries somewhat fuzzy. The rest of the near-Earth asteroids are driven out of the asteroid belt by gravitational interactions with Jupiter.[4] [25]

There are three families of near-Earth asteroids:[4]

- The *Atens*, which have average orbital radii less than one AU[26] and aphelia of more than Earth's perihelion (0.983 AU), placing them usually inside the orbit of Earth.
- The *Apollos*, which have average orbital radii more than that of the Earth and perihelia less than Earth's aphelion (1.017 AU).
- The *Amors*, which have average orbital radii in between the orbits of Earth and Mars and perihelia slightly outside Earth's orbit (1.017–1.3 AU). Amors often cross the orbit of Mars, but they do not cross the orbit of Earth.

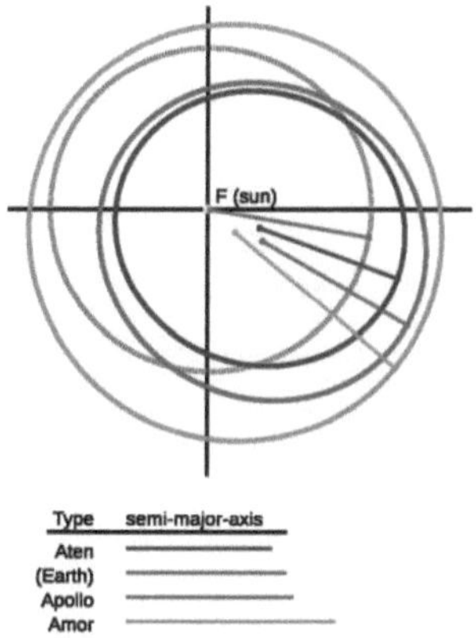

The three families of near-Earth asteroids

Many Atens and all Apollos have orbits that cross (though not necessarily intersect) that of the Earth, so they are a threat to impact the Earth on their current orbits. Amors do not cross the Earth's orbit and are not immediate impact threats. However, their orbits may evolve into Earth-crossing orbits in the future.

Also sometimes used is the Arjuna asteroid classification, for asteroids with extremely Earth-like orbits.[27]

Near-Earth comets

As of May 2010, 84 near-Earth comets have been discovered.[18] Although no impact of a comet in Earth´s history has been conclusively confirmed, the Tunguska event may have been caused by a fragment of Comet Encke.[28] Cometary fragmenting may also be responsible for some impacts from near-Earth objects.

These near-Earth objects were probably derived from the Kuiper belt, beyond the orbit of Neptune.

Impact rate

Objects with diameters of 5-10 m impact the Earth's atmosphere approximately once per year, with as much energy as the atomic bomb dropped on Hiroshima, approximately 15 kilotonnes of TNT. These ordinarily explode in the upper atmosphere, and most or all of the solids are vaporized.[29] Every 2000–3000 years NEAs produce explosions comparable to the one observed at Tunguska in 1908.[30] Objects with a diameter of one kilometer hit the Earth an average of twice every million year interval.[4] Large collisions with five kilometer objects happen approximately once every ten million years.

The rate of impacts of objects of at least 1 km in diameter is estimated as 2 per million years. Assuming that this rate will continue for the next billion years, there exist at least 2,000 objects of diameter greater than 1 km that will eventually hit Earth. However, most of these are not yet considered potentially hazardous objects because they are currently orbiting between Mars and Jupiter. Eventually they will change orbits and become NEOs. Objects spend on average a few million years as NEOs before hitting the Sun, being ejected from the Solar System, or (for a small proportion) hitting a planet.[31]

Historic impacts

The general acceptance of the Alvarez hypothesis, explaining the Cretaceous–Tertiary extinction event as the result of a large object impact event, raised the awareness of the possibility of future Earth impacts with other objects that cross the Earth's orbit.[2]

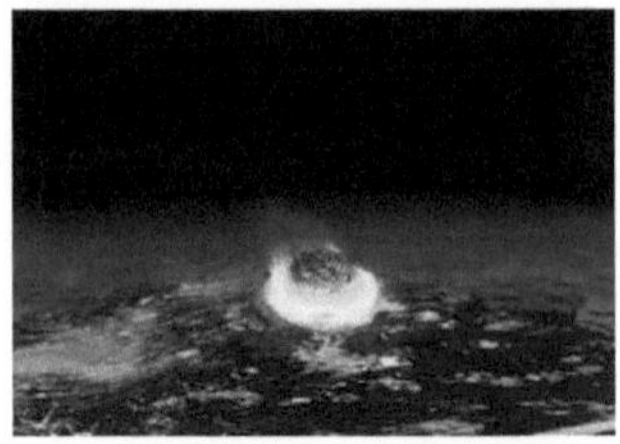

Illustration of the impact of an asteroid a few kilometers across. Such impacts are expected to occur less often than every 100 million years.

1908 Tunguska event

It is now commonly believed that on 30 June 1908 a stony asteroid exploded over Tunguska with the energy of the explosion of 10 megatons of TNT. The explosion occurred at a height of 8.5 kilometers. The object that caused the explosion has been estimated to have had a diameter of 45–70 meters.[32]

1979 Vela Incident

A 22 September 1979 event recorded as occurring near the junction of the South Atlantic and the Indian Ocean was possibly a low-yield nuclear test, but was also initially thought to have been caused by the possible impact of an extraterrestrial object. The event, which became known as the Vela Incident, was identified by a U.S. Vela defence satellite in Earth orbit. The event alarm triggered multi-year investigations by several organizations which could not conclusively determine if the explosion was of nuclear or non-nuclear origin.

2002 Eastern Mediterranean event

On 6 June 2002 an object with an estimated diameter of 10 meters collided with Earth. The collision occurred over the Mediterranean Sea, between Greece and Libya, at approximately 34°N 21°E and the object exploded in mid-air. The energy released was estimated (from infrasound measurements) to be equivalent to 26 kilotons of TNT, comparable to a small nuclear weapon.[33]

2008 Sudan event

On 6 October 2008, scientists calculated that a small near-Earth asteroid, 2008 TC_3, just sighted that night, would impact the Earth on 7 October over Sudan, at 0246 UTC, 5:46 local time.[34] [35] The asteroid arrived as predicted.[36] [37] This is the first time that an asteroid impact on Earth has been accurately predicted. However, no reports of the actual impact have so far been published since it occurred in a very sparsely populated area.[38]

2009 Indonesia event

A large fireball was observed in the skies near Bone, Indonesia on October 8, 2009. This was thought to be caused by an asteroid approximately 10 meters in diameter. The fireball contained an estimated energy of 50 kilotons of TNT, or about twice the Nagasaki atomic bomb. No injuries were reported.[39]

Close approaches

Flyby of asteroid 2004 FH. The other object that flashes by is an artificial satellite.

On August 10, 1972 a meteor that became known as The Great Daylight 1972 Fireball was witnessed by many people moving north over the Rocky Mountains from the U.S. Southwest to Canada. It was an Earth-grazing meteoroid that passed within 57 kilometers (about 34 miles) of the Earth's surface. It was filmed by a tourist at the Grand Teton National Park in Wyoming with an 8-millimeter color movie camera.[40]

On March 23, 1989 the 300-meter (1,000-foot) diameter Apollo asteroid 4581 Asclepius (1989 FC) missed the Earth by **unknown operator: u','** kilometers (**unknown operator: u'strong'unknown operator: u','**mi) passing through the exact position where the Earth was only 6 hours before. If the asteroid had impacted it would have created the largest explosion in recorded history, twelve times more powerful than the Tsar Bomba, the most powerful nuclear bomb ever exploded by man. It attracted widespread attention as early calculations had its passage being as close as **unknown operator: u','unknown operator: u','unknown operator: u','** (**unknown operator: u'strong'unknown operator: u','**mi) from the Earth, with large uncertainties that allowed for the possibility of it striking the Earth.[41]

On March 18, 2004, LINEAR announced a 30-meter asteroid, 2004 FH, which would pass the Earth that day at only **unknown operator: u','unknown operator: u','unknown operator: u','** (**unknown operator: u'strong'unknown operator: u','**mi), about one-tenth the distance to the Moon, and the closest miss ever noticed. They estimated that similar-sized asteroids come as close about every two years.[42]

On March 31, 2004, two weeks after 2004 FH, meteoroid 2004 FU_{162} set a new record for closest recorded approach, passing Earth only **unknown operator: u','unknown operator: u','unknown operator: u','** (**unknown operator: u'strong'unknown operator: u','**mi) away (about one-sixtieth of the distance to the Moon). Because it was very small (6 meters/20 feet), FU_{162} was detected only hours before its closest approach. If it had collided with Earth, it probably would have harmlessly disintegrated in the atmosphere.

On March 2, 2009, near-Earth asteroid 2009 DD45 flew by Earth at about 13:40 UT. The estimated distance from Earth was **unknown operator: u','unknown operator: u','unknown operator: u','** (**unknown operator: u'strong'unknown operator: u','**mi), approximately twice the height of a geostationary communications satellite. The estimated size of the space rock was about 35 meters (115 feet) wide.[43]

On January 13, 2010 at 12:46 UT, near-Earth asteroid 2010 AL30[44] passed at about **unknown operator: u','unknown operator: u','unknown operator: u','** (**unknown operator: u'strong'unknown operator: u','**mi). It was approximately 10–15 m (33–49 ft) wide. If 2010 AL30 had entered the Earth's atmosphere, it would have created an air burst equivalent to between 50 kT and 100 kT (kilotons of TNT). The Hiroshima "Little Boy" atom bomb had a yield between 13-18kT.[45]

On June 28, 2011 an asteroid designated 2011 MD, estimated at 5–20 m (16–66 ft) in diameter, passed within **unknown operator: u','unknown operator: u','unknown operator: u','** (**unknown operator: u'strong'unknown operator: u','**mi) of the Earth, passing over the Atlantic Ocean.[46]

On November 8, 2011 2005 YU55 (at about 400m diameter) passed within 324600 km (mi) (0.85 lunar distances) from Earth.

Future impacts

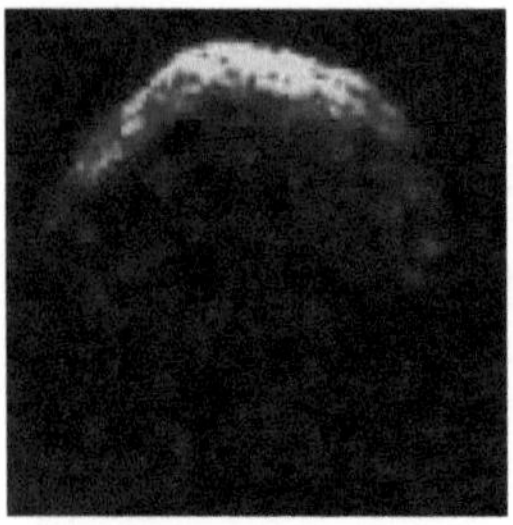

Radar image of asteroid 1950 DA.

Although there have been a few false alarms, a number of objects have been known to be threats to the Earth. (89959) 2002 NT7 was the first asteroid with a positive rating on the Palermo Technical Impact Hazard Scale, with approximately one in a million on a potential impact date of February 1, 2019. It is now known that on January 13, 2019, 2002 NT_7 will safely pass 0.4078 AU (km; mi) from the Earth.[47]

Asteroid (29075) 1950 DA was lost after its discovery in 1950 since not enough observations were made to allow plotting of its orbit, and then rediscovered on December 31, 2000. The chance it will impact Earth on March 16, 2880 during its close approach has been estimated as 1 in 300. This chance of impact for such a large object is roughly 50% greater than that for all other such objects combined between now and 2880.[48] It has a diameter of about a kilometer (0.6 miles).

Only the asteroids 99942 Apophis (provisionally known as 2004 MN4) and (144898) 2004 VD_{17} have briefly had above-normal rankings on the Torino Scale.

Projects to minimize the threat

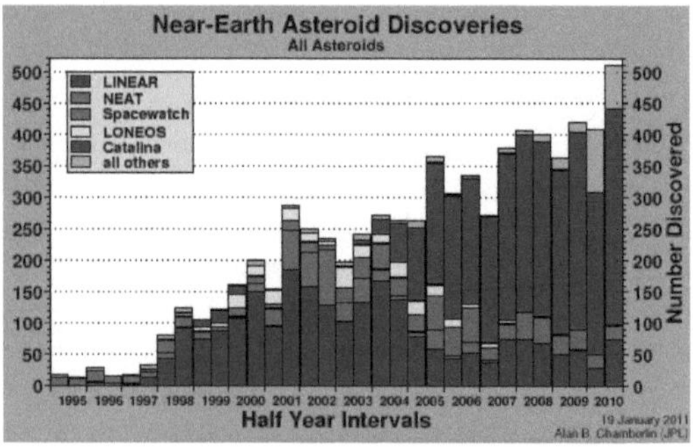

Number of NEOs detected by various projects.

Several surveys have undertaken "Spaceguard" activities (an umbrella term), including Lincoln Near-Earth Asteroid Research (LINEAR), Spacewatch, Near-Earth Asteroid Tracking (NEAT), Lowell Observatory Near-Earth-Object Search (LONEOS), Catalina Sky Survey, Campo Imperatore Near-Earth Objects Survey (CINEOS), Japanese Spaceguard Association, and Asiago-DLR Asteroid Survey. In 1998, the United States Congress mandated the Spaceguard Survey – detection of 90% of near-earth asteroids over 1 km diameter (which threaten global devastation) by 2008. This could be extended by the George E. Brown, Jr. Near-Earth Object Survey Act, which calls for NASA to detect 90 percent of NEOs with diameters of 140 meters or greater by 2020.[49]

As of 2011, 911 of the largest (>1 km diameter) near-Earth asteroids have been found, with some 70 still missing.[50]

See also

- List of Earth-crossing minor planets
- 6Q0B44E, in Earth orbit with a period of 80 days
- J002E3, probably the 3rd stage of Apollo 12
- BRAMS (Belgian RAdio Meteor Stations); study of meteor population[51]
- Bolide
- Co-orbital configuration
- Euronear
- Mission Marco Polo
- NEODyS
- Orbit@home
- Space debris
- 2010 SO16
- Asteroid mining
- Potentially hazardous object

References

[1] Glossary of Astronomical Terms (http://www.astunit.com/tutorials/glossary.htm#N)

[2] Richard Monastersky (March 1, 1997). "The Call of Catastrophes" (http://www.sciencenews.org/pages/sn_arc97/75th/rm_essay.htm). Science News Online. . Retrieved 2007-10-23.

[3] Krajik, K. (2007) *Killer Space Rocks*. Popular Science magazine, September. pg 72-73 (http://books.google.com/books?id=3AClY8pMg-EC&lpg=PA56&dq=Popular Science Sept 2007&pg=PA72#v=onepage&q&f=false).

[4] A. Morbidelli, W. F. Bottke Jr., Ch. Froeschlé, P. Michel (January 2002). W. F. Bottke Jr., A. Cellino, P. Paolicchi, and R. P. Binzel. ed. "Origin and Evolution of Near-Earth Objects" (http://www.boulder.swri.edu/~bottke/Reprints/Morbidelli-etal_2002_AstIII_NEOs.pdf) (PDF). *Asteroids III* (University of Arizona Press): 409–422. .

[5] Clark R. Chapman (May 2004). "The hazard of near-Earth asteroid impacts on earth". *Earth and Planetary Science Letters* **222** (1): 1–15. Bibcode 2004E&PSL.222....1C. doi:10.1016/j.epsl.2004.03.004.

[6] Shiga, David (2006-06-27). "New telescope will hunt dangerous asteroids" (http://www.newscientist.com/article/dn9403). New Scientist. . Retrieved 2008-11-15.

[7] "Unusual Minor Planets" (http://www.minorplanetcenter.org/iau/lists/Unusual.html). Minor Planet Center. 2008-10-21. . Retrieved 2010-10-25.

[8] "Potentially Hazard Asteroids" (http://neo.jpl.nasa.gov/neo/pha.html). NASA/JPL Near-Earth Object Program Office. . Retrieved 2011-05-05.

[9] http://neo.jpl.nasa.gov/neo/groups.html

[10] NEO Earth Close Approaches (http://neo.jpl.nasa.gov/ca/) at NASA/JPL Near-Earth Object Program Office

[11] Dan Vergano (February 2, 2007). "Near-Earth asteroids could be 'steppingstones to Mars'" (http://www.usatoday.com/tech/science/space/2007-02-12-asteroid_x.htm). USA Today. . Retrieved 2007-10-22.

[12] Rui Xu, Pingyuan Cui, Dong Qiao and Enjie Luan (18 March 2007). "Design and optimization of trajectory to Near-Earth asteroid for sample return mission using gravity assists". *Advances in Space Research* **40** (2): 200–225. Bibcode 2007AdSpR..40..220X. doi:10.1016/j.asr.2007.03.025.

[13] Donald Savage and Michael Buckley (January 31, 2001). "NEAR Mission Completes Main Task, Now Will Go Where No Spacecraft Has Gone Before" (http://nssdc.gsfc.nasa.gov/planetary/news/near_descent_pr_20010131.html). National Aeronautics and Space Administration. . Retrieved 2007-10-22.

[14] Don Yeomans (August 11, 2005). "Hayabusa's Contributions Toward Understanding the Earth's Neighborhood" (http://neo.jpl.nasa.gov/missions/hayabusa.html). National Aeronautics and Space Administration. . Retrieved 2007-10-22.

[15] "The flux of small near-Earth objects colliding with the Earth" by P. Brown et al., Nature, 420, pp. 294–6, November 2002.

[16] "Current Impact Risks" (http://neo.jpl.nasa.gov/risk). . Retrieved 2010-05-28.

[17] NASA's *Near Earth Object Program* (http://neo.jpl.nasa.gov/1950da/)

[18] NEO Discovery Statistics (http://neo.jpl.nasa.gov/stats/)

[19] The Torino Impact Hazard Scale (http://neo.jpl.nasa.gov/torino_scale.html)

[20] Dr. David R. Williams (September 13, 2006). "Near Earth Object Fact Sheet" (http://nssdc.gsfc.nasa.gov/planetary/factsheet/neofact.html). National Aeronautics and Space Administration. . Retrieved 2007-10-22.

[21] Jane Platt (January 12, 2000). "Asteroid Population Count Slashed" (http://www.jpl.nasa.gov/releases/2000/neat.html). National Aeronautics and Space Administration. . Retrieved 2007-10-22.

[22] David Rabinowitz, Eleanor Helin, Kenneth Lawrence and Steven Pravdo (13 January 2000). "A reduced estimate of the number of kilometer-sized near-Earth asteroids" (http://www.nature.com/nature/journal/v403/n6766/full/403165a0.html). *Nature* **403** (6766): 165–166. Bibcode 2000Natur.403..165R. doi:10.1038/35003128. PMID 10646594. . Retrieved 2007-10-22.

[23] D.F. Lupishko and T.A. Lupishko (May 2001). "On the Origins of Earth-Approaching Asteroids". *Solar System Research* **35** (3): 227–233. Bibcode 2001SoSyR..35..227L. doi:10.1023/A:1010431023010.

[24] A. Morbidelli, D. Vokrouhlický (May 2003). "The Yarkovsky-driven origin of near-Earth asteroids". *Icarus* **163** (1): 120–134. Bibcode 2003Icar..163..120M. doi:10.1016/S0019-1035(03)00047-2.

[25] D.F. Lupishko, M. di Martino and T.A. Lupishko (September 2000). "What the physical properties of near-Earth asteroids tell us about sources of their origin?". *Kinematika i Fizika Nebesnykh Tel Supplimen* **3** (3): 213–216. Bibcode 2000KFNTS...3..213L.

[26] The distance from the Earth to the Sun

[27] Ron Cowen (February 20, 1993). "Near-Earth asteroids: class consciousness – new asteroids identified" (http://findarticles.com/p/articles/mi_m1200/is_n8_v143/ai_13526583). Science News. . Retrieved 2007-10-23.

[28] "The Tunguska object – A fragment of Comet Encke" (http://adsabs.harvard.edu/abs/1978BAICz..29..129K). Astronomical Institutes of Czechoslovakia. . Retrieved 2007-10-15.

[29] Clark R. Chapman & David Morrison (6 January 1994). "Impacts on the Earth by asteroids and comets: assessing the hazard". *Nature* **367** (6458): 33–40. Bibcode 1994Natur.367...33C. doi:10.1038/367033a0.

[30] EARTH IN THE COSMIC SHOOTING GALLERY, D. J. Asher et al. page 2 (http://www.arm.ac.uk/preprints/455.pdf)

[31] Morbidelli, A., W. F. Bottke, Ch. Froeschle, and P. Michel. 2002. Origin and evolution of near-Earth objects. (http://www.boulder.swri.edu/~bottke/Reprints/Morbidelli-etal_2002_AstIII_NEOs.pdf) In Asteroids III, (W. F. Bottke, A. Cellino, P. Paolicchi, R. Binzel, Eds). U. Arizona Press, 409-422.

[32] Christopher F. Chyba, Paul J. Thomas & Kevin J. Zahnle (January 7, 1993). "The 1908 Tunguska explosion: atmospheric disruption of a stony asteroid". *Nature* **361** (6407): 40–44. Bibcode 1993Natur.361...40C. doi:10.1038/361040a0.

[33] P. Brown, R.E. Spalding, D.O. ReVelle, E. Tagliaferri and S.P. Worden (21 November 2002). "The flux of small near-Earth objects colliding with the [[Earth (http://www.astro.uwo.ca/~pbrown/documents/flux-final.pdf)]"] (PDF). *Nature* **420** (6913): 294–296. doi:10.1038/nature01238. PMID 12447433. . Retrieved 2007-10-23.

[34] Don Yeomans (October 6, 2008). "Small Asteroid Predicted to Cause Brilliant Fireball over Northern Sudan" (http://neo.jpl.nasa.gov/news/news159.html). NASA/JPL Near-Earth Object Program Office. . Retrieved 2008-10-09.

[35] Richard A. Kerr (6 October 2008). "FLASH! Meteor to Explode Tonight" (http://sciencenow.sciencemag.org/cgi/content/full/2008/1006/2). ScienceNOW Daily News. . Retrieved 2008-10-09.

[36] Don Yeomans (October 7, 2008). "Impact of Asteroid 2008 TC3 Confirmed" (http://neo.jpl.nasa.gov/news/news160.html). NASA/JPL Near-Earth Object Program Office. . Retrieved 2008-10-09.

[37] Richard A. Kerr (8 October 2008). "Asteroid Watchers Score a Hit" (http://sciencenow.sciencemag.org/cgi/content/full/2008/1008/1). ScienceNOW Daily News. . Retrieved 2008-10-09.

[38] Little Asteroid Makes a Big Splash (http://www.skyandtelescope.com/community/skyblog/newsblog/30686199.html) Sky and Telescope, October 9, 2008.

[39] http://neo.jpl.nasa.gov/news/news165.html

[40] Grand Teton Meteor Video (http://www.youtube.com/watch?v=7M8LQ7_hWtE), Youtube

[41] Brian G. Marsden (1998-03-29). "HOW THE ASTEROID STORY HIT: AN ASTRONOMER REVEALS HOW A DISCOVERY SPUN OUT OF CONTROL" (http://www.cfa.harvard.edu/iau/pressinfo/1997XF11Globe.html). The Boston Globe. . Retrieved 2007-10-23.

[42] Steven R. Chesley and Paul W. Chodas (March 17, 2004). "Recently Discovered Near-Earth Asteroid Makes Record-breaking Approach to Earth" (http://neo.jpl.nasa.gov/news/news142.html). National Aeronautics and Space Administration. . Retrieved 2007-10-23.

[43] Yahoo News – 20090302 – Asteroid Flies Past Earth (http://news.yahoo.com/s/space/20090302/sc_space/asteroidfliespastearth)

[44] Small Asteroid 2010 AL30 Will Fly Past The Earth (http://neo.jpl.nasa.gov/news/news167.html). NASA/JPL Near-Earth Object Program, January 12, 2010.

[45] Near-Earth Object 2010 AL30 (http://epod.usra.edu/blog/2010/03/nearearth-object-2010-al30.html). NASA Earth Science Picture of the Day March 06, 2010.

[46] "Asteroid soars over Atlantic Ocean" (http://www.abc.net.au/news/stories/2011/06/28/3255543.htm). Australian Broadcasting Corporation. June 28, 2011. . Retrieved 2011-07-03.

[47] "JPL Close-Approach Data: 89959 (2002 NT7)" (http://ssd.jpl.nasa.gov/sbdb.cgi?sstr=2002NT7;cad=1#cad). 2011-09-12 last obs (arc=57 years). . Retrieved 2011-11-04.

[48] Giorgini, J. D.; Ostro, S. J.; Benner, L. A. M.; Chodas, P. W.; Chesley, S. R.; Hudson, R. S.; Nolan, M. C.; Klemola, A. R. et al. (April 5, 2002). "Asteroid 1950 DA's Encounter with Earth in 2880: Physical Limits of Collision Probability Prediction" (http://neo.jpl.nasa.gov/1950da/1950da.pdf) (PDF). *Science* **296** (5565): 132–136. Bibcode 2002Sci...296..132G. doi:10.1126/science.1068191. PMID 11935024. . Retrieved January 11, 2010.

[49] National Academy of Sciences Jan. 23, 2010 (http://www8.nationalacademies.org/onpinews/newsitem.aspx?RecordID=12842) Defending Planet Earth: Near-Earth Object Surveys and Hazard Mitigation Strategies: Final Report. Washington, DC: The National Academies Press. Book available at: http://books.nap.edu/catalog.php?record_id=12842.

[50] http://www.nasa.gov/mission_pages/WISE/multimedia/gallery/neowise/pia14734.html
[51] Belgian RAdio Meteor Stations (http://brams.aeronomie.be/)

External links

- Interactive Earth damage impact calculator from NEO (http://www.purdue.edu/impactearth)
- The Jet Propulsion Laboratory's NEO website (http://neo.jpl.nasa.gov) / NEO Earth Close Approaches (http://neo.jpl.nasa.gov/ca/)
- TECA Table of next close approaches to the Earth (http://www.brera.mi.astro.it/sormano/teca.html)
- SAEL Small Asteroids Encounter List (http://www.brera.mi.astro.it/sormano/sael.html)
- MBPL Minor Body Priority List (http://www.brera.mi.astro.it/sormano/mbpl.html)
- Spaceguard UK (http://www.spaceguarduk.com/)
- House Holds Hearing On NEO Threat (http://starrymirror.com/neo hearing1109.htm)
- The UK NEO Information Centre (http://www.spacecentre.co.uk/Page.aspx/6/NEAR_EARTH_OBJECTS/)
- Earth in the Cosmic Shooting Gallery (http://www.arm.ac.uk/preprints/455.pdf) Paper published 2005
- Near Earth Asteroids (http://www.astronomycast.com/astronomy/episode-29-asteroids-make-bad-neighbors/) Astronomy Cast episode #29, includes full transcript.
- Upcoming Close Approaches (<0.10 A.U.) of Near-Earth Objects to Earth (http://members.shaw.ca/andrewlowe/PHA-SOON.HTM#Earth)
- Catalogue of the Solar System Small Bodies Orbital Evolution (http://smallbodies.ru/en/)
- NASA Space Telescope Finds Fewer Asteroids Near Earth (http://www.nasa.gov/mission_pages/WISE/news/wise20110929.html) (09.29.11)

Centaur_(minor_planet)

Centaurs are an unstable orbital class of minor planets that behave with characteristics of both asteroids and comets. They are named after the mythological race of beings, centaurs, which were a mixture of horse and human. Centaurs have transient orbits that cross or have crossed the orbits of one or more of the giant planets, and have dynamic lifetimes of a few million years.[1] It has been estimated that there are around 44,000 centaurs in the Solar System with diameters larger than 1 km.[1]

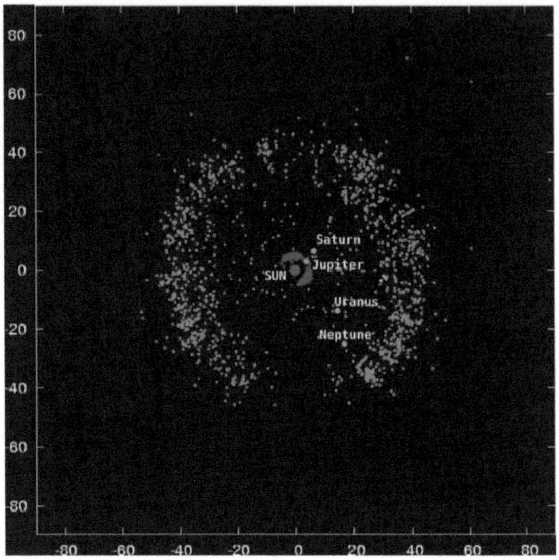

Positions of known outer Solar System objects. The centaurs are those objects (in orange) that lie generally **inwards** of the Kuiper belt (in green) and outside the Jupiter Trojans (pink).

The first centaur-like object to be discovered was 944 Hidalgo in 1920. However, they were not recognized as a distinct population until the discovery of 2060 Chiron in 1977. The largest known centaur is 10199 Chariklo, discovered in 1997, which at 260 km in diameter is as big as a mid-sized main-belt asteroid.

No centaur has been photographed up close, although there is evidence that Saturn's moon Phoebe, imaged by the Cassini probe in 2004, may be a captured centaur. In addition, the Hubble Space Telescope has gleaned some information about the surface features of 8405 Asbolus.

As of 2008, three centaurs have been found to display cometary comas: Chiron, 60558 Echeclus, and 166P/NEAT. Chiron and Echeclus are therefore classified as both asteroids and comets. Other centaurs such as 52872 Okyrhoe are suspected of showing cometary activity. Any centaur that is perturbed close enough to the Sun is expected to become a comet.

Classification

The generic definition of a centaur is a small body that orbits the Sun between Jupiter and Neptune and crosses the orbits of one or more of the giant planets. Due to the inherent long-term instability of orbits in this region, even centaurs such as 2000 GM_{137} and 2001 XZ_{255}, which do not currently cross the orbit of any planet, are in gradually changing orbits that will be perturbed until they start to cross the orbit of one or more of the giant planets.[1]

However, different institutions have different criteria for classifying borderline objects, based on particular values of their orbital elements:

- The Minor Planet Center (MPC) defines centaurs as having a perihelion beyond the orbit of Jupiter and a semi-major axis less than that of Neptune.[2]
- The Jet Propulsion Laboratory (JPL) similarly defines centaurs as having a semi-major axis, a, between those of Jupiter and Neptune (5.5 AU < a < 30.1 AU).[3]

- In contrast, the Deep Ecliptic Survey (DES) defines centaurs using a dynamical classification scheme. These classifications are based on the simulated change in behavior of the present orbit when extended over 10 million years. The DES defines centaurs as non-resonant objects whose instantaneous (osculating) perihelia are less than the osculating semi-major axis of Neptune at any time during the simulation. This definition is intended to be synonymous with planet-crossing orbits and to suggest comparatively short lifetimes in the current orbit.[4]

The collection *The Solar System Beyond Neptune* (2008) uses the traditional definition of centaurs, limited to semi-major axes smaller than that of Neptune, classifying the objects on unstable orbits beyond this limit as members of the scattered disk.[5] Yet, other astronomers still prefer to define centaurs as objects that are non-resonant with a perihelion inside the orbit of Neptune that can be shown to likely cross the Hill sphere of a gas giant within the next 10 million years.[6] Thus centaurs can be thought of as inward scattered objects that interact more aggressively and scatter more quickly than typical scattered disc objects.

The JPL Small-Body Database lists 183 centaurs.[7] There are an additional 40 trans-Neptunian objects with a semi-major axis further than Neptune ($a > 30.1$ AU) and perihelion closer than the orbit of Uranus ($q < 19$ AU).[7] The Committee on Small Body Nomenclature of the International Astronomical Union has not formally weighed in on either side of the debate. Instead, it has adopted the following naming convention for such objects: befitting their centaur-like transitional orbits between TNOs and comets, "objects on unstable, non-resonant, giant-planet-crossing orbits with semimajor axes greater than Neptune's" are to be named for other hybrid and shape-shifting mythical creatures. Thus far, only the binary objects Ceto and Phorcys and Typhon and Echidna have been named according to the new policy.[8]

Other objects caught between these differences in classification methods include (44594) 1999 OX_3, which has a semi-major axis of 32 AU but crosses the orbits of both Uranus and Neptune. Among the inner centaurs, 2005 VD, with a perihelion distance very near Jupiter, is listed as a centaur by both JPL and DES.

Orbits

Distribution

The diagram at right illustrates the orbits of all known centaurs in relation to the orbits of the planets. For selected objects, the eccentricity of the orbits is represented by red segments (extending from perihelion to aphelion).

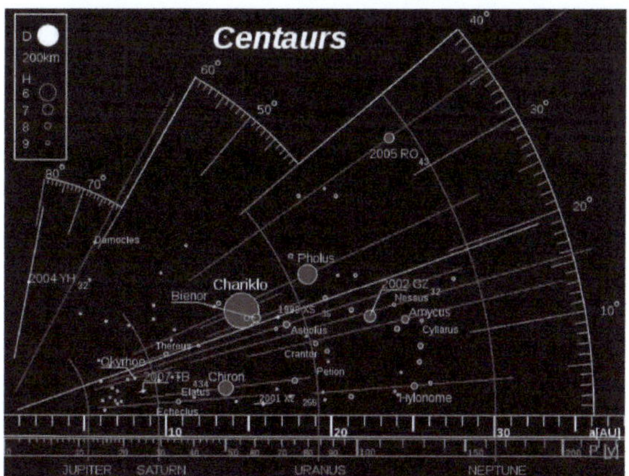

Orbits of known centaurs[9]

Centaurs' orbits are characterised by a wide range of eccentricity, from highly eccentric (Pholus, Asbolus, Amicus, Nessus) to more circular (Chariklo and the *Saturn-crossers*: Thereus, Okyrhoe).

To illustrate the range of the orbits' parameters, a few objects with very unusual orbits are plotted in yellow on the diagram:

- 1999 XS_{35} (Apollo asteroid) follows an extremely eccentric orbit (e=0.947), leading it from inside Earth's orbit (0.94 AU) to well beyond Neptune (>34 AU)

- 2007 TB_{434} follows a quasi-circular orbit (e<0.026)
- 2001 XZ_{255} has the lowest inclination (i<3°).
- Damocles is among a few centaurs on orbits with extreme inclination (prograde i>70°, e.g. 2007 DA_{61}, 2004 YH_{32}, retrograde i<120° e.g. 2005 JT_{50}; not shown)
- 2004 YH_{32} follows such a highly inclined orbit (nearly 80°) that, while it crosses from the distance of the asteroid belt from the Sun to past the distance of Saturn, it does not even cross Jupiter relative to the plane of Jupiter's orbit.

A dozen known centaurs, including Dioretsa ("asteroid" spelled backwards), follow retrograde orbits.

Changing orbits

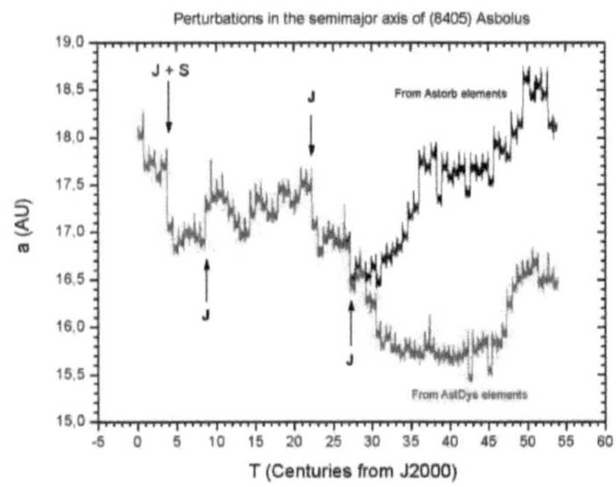

The semi-major axis of Asbolus during the next 5500 years. After the Jupiter encounter of year 4713 (27 Cy in the future) the two orbits become substantially divergent.[10]

Since the centaurs cross the orbits of the giant planets and are not protected by orbital resonances, their orbits are unstable within a timescale of 10^6–10^7 years.[11] For example, 55576 Amycus is in an unstable orbit near the 3:4 resonance of Uranus.[1] Dynamical studies of their orbits indicate that centaurs are probably an intermediate orbital state of objects transitioning from the Kuiper belt to the Jupiter family of short-period comets. Objects may be perturbed from the Kuiper belt, whereupon they become Neptune-crossing and interact gravitationally with that planet (see theories of origin). They then become classed as centaurs, but their orbits are chaotic, evolving relatively rapidly as the centaur makes repeated close approaches to one or more of the outer planets. Some centaurs will evolve into Jupiter-crossing orbits whereupon their perihelia may become reduced into the inner Solar System and they may be reclassified as active comets in the Jupiter family if they display cometary activity. Centaurs will thus ultimately collide with the Sun or a planet or else they may be ejected into interstellar space after a close approach to one of the planets, particularly Jupiter.

Physical characteristics

The relatively small size of centaurs precludes surface observations, but colour indices and spectra can indicate possible surface composition and can provide insight into the origin of the bodies.[11]

Colours

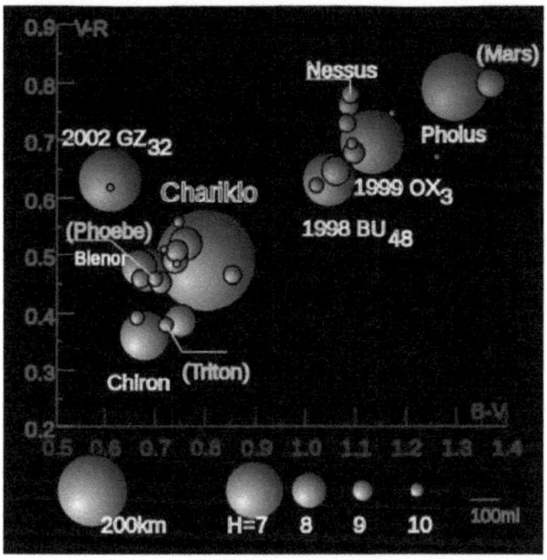

Colour distribution of centaurs

Centaurs display a puzzling diversity of colour that challenges any simple model of surface composition.[12] In the side-diagram, the colour indices are measures of apparent magnitude of an object through blue (B), visible (V) *i.e.* green-yellow and red (R) filters. The diagram illustrates these differences (in enhanced colour) for all centaurs with known colour indices. For reference, two moons: Triton and Phoebe, and planet Mars are plotted (yellow labels, size not to scale).

Centaurs appear to be grouped into two classes:

- very red, for example 5145 Pholus
- blue (or blue-grey, according to some authors), for example 2060 Chiron

There are numerous theories to explain this colour difference, but they can be divided broadly into two categories:

- The colour difference results from a difference in the origin and/or composition of the centaur (see origin below)
- The colour difference reflects a different level of space-weathering from radiation and/or cometary activity.

As examples of the second category, the reddish colour of Pholus has been explained as a possible mantle of irradiated red organics, whereas Chiron has instead had its ice exposed due to its periodic cometary activity, giving it a blue/grey index. The correlation with activity and color is not certain, however, as the active centaurs span the range of colors from blue (Chiron) to red (166P/NEAT).[13] Alternatively, Pholus may have been only recently expelled from the Kuiper belt, so that surface transformation processes have not yet taken place.

A. Delsanti *et al.* suggest multiple competing processes: reddening by the radiation, and blushing by collisions.[14][15]

Spectra

The interpretation of spectra is often ambiguous, related to particle sizes and other factors, but the spectra offer an insight into surface composition. As with the colours, the observed spectra can fit a number of models of the surface.

Water ice signatures have been confirmed on a number of centaurs[11] (including 2060 Chiron, 10199 Chariklo and 5145 Pholus). In addition to the water ice signature, a number of other models have been put forward:

- Chariklo's surface has been suggested to be a mixture of tholins (like those detected on Titan and Triton) with amorphous carbon.
- Pholus has been suggested to be covered by a mixture of Titan-like tholins, carbon black, olivine[16] and methanol ice.
- The surface of 52872 Okyrhoe has been suggested to be a mixture of kerogens, olivines and small percentage of water ice.

- 8405 Asbolus has been suggested to be a mixture of 15% Triton-like tholins, 8% Titan-like tholin, 37% amorphous carbon and 40% ice tholin.

Chiron, the only centaur with known cometary activity, appears to be the most complex. The spectra observed vary depending on the period of the observation. Water ice signature was detected during a period of low activity and disappeared during high activity.[18] [19] [20]

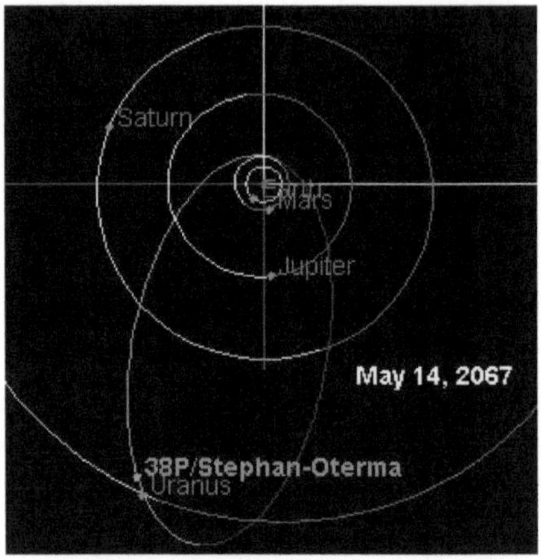

Comet 38P exhibits centaur-like behavior by making close approaches to Jupiter, Saturn, and Uranus between 1982 and 2067.[17]

Similarities to comets

Observations of Chiron in 1988 and 1989 near its perihelion found it to display a coma (a cloud of gas and dust evaporating from its surface). It is thus now officially classified as both a comet and an asteroid, although it is far larger than a typical comet and there is some lingering controversy. Other centaurs are being monitored for comet-like activity: so far two, 60558 Echeclus, and 166P/NEAT have shown such behavior. 166P/NEAT was discovered while it exhibited a coma, and so is classified as a comet, though its orbit is that of a centaur. 60558 Echeclus was discovered without a coma but recently became active,[21] and so it is now accordingly also classified as both a comet and an asteroid.

There is no clear orbital distinction between centaurs and comets. Both 29P/Schwassmann-Wachmann and 39P/Oterma have been referred to as centaurs since they have typical centaur orbits. The comet 39P/Oterma is currently inactive and was seen to be active only before it was perturbed into a centaur orbit by Jupiter in 1963.[22] The faint comet 38P/Stephan-Oterma would probably not show a coma if it had a perihelion distance beyond Jupiter's orbit at 5 AU. By the year 2200, comet 78P/Gehrels will probably migrate outwards into a centaur-like orbit.

Theories of origin

The study of centaur development is rich in recent developments but still hampered by limited physical data. Different models have been put forward for possible origin of centaurs.

Simulations indicate that the orbit of some Kuiper-belt objects can be perturbed, resulting in the object's expulsion so that it becomes a centaur. Scattered disk objects would be dynamically the best candidates[23] for such expulsions, but their colours do not fit the bicoloured nature of the centaurs. Plutinos are a class of Kuiper-belt object that display a similar bicoloured nature, and there are suggestions that not all plutinos' orbits are as stable as initially thought, due to perturbation by Pluto.[24] Further developments are expected with more physical data on KBOs.

Notable centaurs

Well-known centaurs include:

Name	Year	Discoverer	Half-life[1] (forward)	Class
55576 Amycus	2002	NEAT at Palomar	11.1 Myr	UE
10370 Hylonome	1995	Mauna Kea Observatory	6.3 Myr	UN
10199 Chariklo	1997	Spacewatch	10.3 Myr	U
8405 Asbolus	1995	Spacewatch (James V. Scotti)	0.86 Myr	SN
7066 Nessus	1993	Spacewatch (David L. Rabinowitz)	4.9 Myr	SE
5145 Pholus	1992	Spacewatch (David L. Rabinowitz)	1.28 Myr	SN
2060 Chiron	1977	Charles T. Kowal	1.03 Myr	SU

Notes

[1] Horner, J.; Evans, N.W.; Bailey, M. E. (2004). "Simulations of the Population of Centaurs I: The Bulk Statistics". *Monthly Notices of the Royal Astronomical Society* **354** (3): 798–810. arXiv:astro-ph/0407400. Bibcode 2004MNRAS.354..798H. doi:10.1111/j.1365-2966.2004.08240.x.

[2] "Unusual Minor Planets" (http://www.minorplanetcenter.org/iau/lists/Unusual.html). Minor Planet Center. . Retrieved 2010-10-25.

[3] "Orbit Classification (Centaur)" (http://ssd.jpl.nasa.gov/sbdb_help.cgi?class=CEN). JPL Solar System Dynamics. . Retrieved 2008-10-13.

[4] Elliot, J.L.; Kern, Buie, Trilling; et al. (2005). "The Deep Ecliptic Survey: A Search for Kuiper Belt Objects and Centaurs. II. Dynamical Classification, the Kuiper Belt Plane, and the Core Population" (http://www.iop.org/EJ/abstract/1538-3881/129/2/1117). *The Astronomical Journal* **129** (2): 1117–1162. Bibcode 2005AJ....129.1117E. doi:10.1086/427395. . Retrieved 2008-09-22.

[5] B. Gladman, B. Marsden, C. VanLaerhoven (2008). "Nomenclature in the Outer Solar System". *In **The Solar System Beyond Neptune**, ISBN 987-0-8165-2755-7.*

[6] Chaing, Eugene; Buie, Grundy, Holman; et al. (2007). "A Brief History of Transneptunian Space". *Protostars and Planets V, B. Reipurth, D. Jewitt, and K. Keil (eds.), University of Arizona Press, Tucson*: 895–911. arXiv:astro-ph/0601654. Bibcode 2006astro.ph..1654C.

[7] "JPL Small-Body Database Search Engine" (http://ssd.jpl.nasa.gov/sbdb_query.cgi). JPL Solar System Dynamics. . Retrieved 2010-12-27.

[8] Grundy, Will; Stansberry, J.A.; Noll, K; Stephens, D.C.; Trilling, D.E.; Kern, S.D.; Spencer, J.R.; Cruikshank, D.P. et al. (2007). "The orbit, mass, size, albedo, and density of (65489) Ceto/Phorcys: A tidally-evolved binary Centaur". *Icarus* **191** (1): 286–297. arXiv:0704.1523. Bibcode 2007Icar..191..286G. doi:10.1016/j.icarus.2007.04.004.

[9] For the purpose of this diagram, an object is classified as a centaur if its semi-major axis lies between Jupiter and Neptune. Last update: October 2008

[10] "Three clones of Centaur 8405 Asbolus making passes within 450Gm" (http://home.comcast.net/~kpheider/AsbolusClones.txt). . Retrieved 2009-05-02. (Solex 10) (http://chemistry.unina.it/~alvitagl/solex/)

[11] Jewitt, David C.; A. Delsanti (2006). "The Solar System Beyond The Planets". *Solar System Update : Topical and Timely Reviews in Solar System Sciences*. Springer-Praxis Ed.. ISBN 3-540-26056-0. (Preprint version (pdf) (http://www.ifa.hawaii.edu/faculty/jewitt/papers/2006/DJ06.pdf))

[12] M. A. Barucci, A. Doressoundiram, and D. P. Cruikshank, "Physical Characteristics of TNOs and Centaurs" (2003), available on the web (http://www.lesia.obspm.fr/~alaind/TNO/Barucci2003_comet2.pdf) (accessed 3/20/2008)

[13] Bauer, J. M., Fernández, Y. R., & Meech, K. J. 2003. " An Optical Survey of the Active Centaur C/NEAT (2001 T4) (http://www.journals.uchicago.edu/PASP/journal/issues/v115n810/203101/203101.html)", Publication of the Astronomical Society of the Pacific", **115**, 981

[14] Peixinho, N.; Doressoundiram, A.; Delsanti, A.; Boehnhardt, H.; Barucci, M. A.; Belskaya, I. (2003). "Reopening the TNOs Color Controversy: Centaurs Bimodality and TNOs Unimodality". *Astronomy and Astrophysics* **410** (3): L29–L32. arXiv:astro-ph/0309428. Bibcode 2003A&A...410L..29P. doi:10.1051/0004-6361:20031420.

[15] Hainaut & Delsanti (2002) *Color of Minor Bodies in the Outer Solar System* Astronomy & Astrophysics, **389**, 641 datasource (http://www.sc.eso.org/~ohainaut/MBOSS)

[16] A class of Magnesium Iron Silicates $(Mg, Fe)_2SiO_4$, common components of igneous rocks.

[17] "JPL Close-Approach Data: 38P/Stephan-Oterma" (http://ssd.jpl.nasa.gov/sbdb.cgi?sstr=38P;cad=1#cad). 1981-04-04 last obs. . Retrieved 2009-05-07.

[18] Dotto, E; Barucci, M A; De Bergh, C, *Colours and composition of the centaurs*, Earth, Moon, and Planets, **92**, no. 1–4, pp. 157–167. (June 2003)

[19] Luu, Jane X.; Jewitt, David; Trujillo, C. A. (2000). "Water Ice on 2060 Chiron and its Implications for Centaurs and Kuiper Belt Objects". *The Astrophysical Journal* **531** (2): L151–L154. arXiv:astro-ph/0002094. Bibcode 2000ApJ...531L.151L. doi:10.1086/312536. PMID 10688775.

[20] Fernandez, Y. R.; Jewitt, D. C.; Sheppard, S. S. (2002). "Thermal Properties of Centaurs Asbolus and Chiron". *The Astronomical Journal* **123** (2): 1050–1055. arXiv:astro-ph/0111395. Bibcode 2002AJ....123.1050F. doi:10.1086/338436.

[21] Y-J. Choi, P.R. Weissman, and D. Polishook *(60558) 2000 EC_98*, IAU Circ., **8656** (Jan. 2006), 2.

[22] Mazzotta Epifani, E.; Palumbo; Capria; Cremonese;; et al. (2006). "The dust coma of the active Centaur P/2004 A1 (LONEOS): a CO-driven environment?" (http://google.com/search?q=cache:www.aanda.org/articles/aa/ps/2006/48/aa5189-06.ps.gz). *Astronomy & Astrophysics* **460** (3): 935–944. Bibcode 2006A&A...460..935M. doi:10.1051/0004-6361:20065189. . Retrieved 2009-05-08.

[23] for instance, the centaurs could be part of an "inner" scattered disc of objects perturbed inwards from the Kuiper belt (http://www.zanestein.com/page4_1.htm).

[24] Wan, X.-S; Huang, T.-Y. (2001). "The orbit evolution of 32 plutinos over 100 million year". *Astronomy and Astrophysics* **368** (2): 700–705. Bibcode 2001A&A...368..700W. doi:10.1051/0004-6361:20010056.

See also

- Asteroid
- Minor planet
- Dwarf planet

References

External links

- List of Centaurs and Scattered-Disk Objects (http://www.minorplanetcenter.org/iau/lists/Centaurs.html)
- Centaurs from The Encyclopedia of Astrobiology Astronomy and Spaceflight (http://www.daviddarling.info/encyclopedia/C/Centaur.html)
- Horner, Jonathan; Lykawka, Patryk Sofia (2010). "Planetary Trojans – the main source of short period comets?". *International Journal of Astrobiology* **9** (04): 227–234. arXiv:1007.2541. Bibcode 2010IJAsB...9..227H. doi:10.1017/S1473550410000212.

Trojan_(astronomy)

In astronomy, a **Trojan** is a minor planet or natural satellite (moon) that shares an orbit with a larger planet or moon, but does not collide with it because it orbits around one of the two Lagrangian points of stability (Trojan points), L_4 and L_5, which lie approximately 60° ahead of and behind the larger body, respectively. Trojan objects are one type of co-orbital object. In this arrangement, the massive star and the smaller planet orbit about their common barycenter—a location in space where the forces of their mutual gravitational attraction balance each other out. A much smaller mass located at one of the Lagrange points is subject to a combined gravitational force that acts through this barycenter. As a consequence, the mass can follow a circular orbit around this point with the same period as the planet, and the arrangement can remain stable over time.[1]

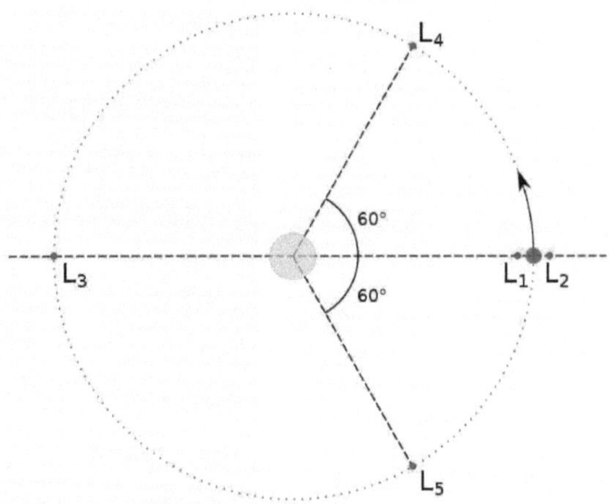

Trojan points are the points labelled L_4 and L_5, highlighted in red, on the orbital path of the secondary object (blue), around the primary object (yellow).

Trojan asteroids are asteroids that reside in a Trojan point of a planet. A Trojan moon is a moon residing at the Trojan point of another (larger) moon. Trojan planets are theoretical planets that reside at Trojan points of other planets.

Saturn has the most known Trojan satellites: Saturn's moon Tethys has two Trojan moons (Telesto and Calypso), and Dione also has two Trojan moons (Helene and Polydeuces).

In 2011, NASA announced the discovery of the first known Earth Trojan.[2]

Trojan asteroids

In 1772 the French mathematician and astronomer Joseph-Louis Lagrange predicted the existence and location of two groups of small bodies located near a pair of gravitationally stable points along Jupiter's orbit. The term originally referred to the Trojan asteroids orbiting around Jupiter's Lagrangian points, which are by convention named after figures from the Trojan War of Greek mythology. By convention, the asteroids orbiting Jupiter's L_4 point are named after the heroes from the Greek side of the war, while those at L_5 are from the Trojan side. The two exceptions, the Greek-themed 617 Patroclus and the Trojan-themed 624 Hektor, were actually assigned to the wrong sides.[3] Astronomers estimate that the Jupiter Trojans are comparable in number to the asteroids of the asteroid belt.[4]

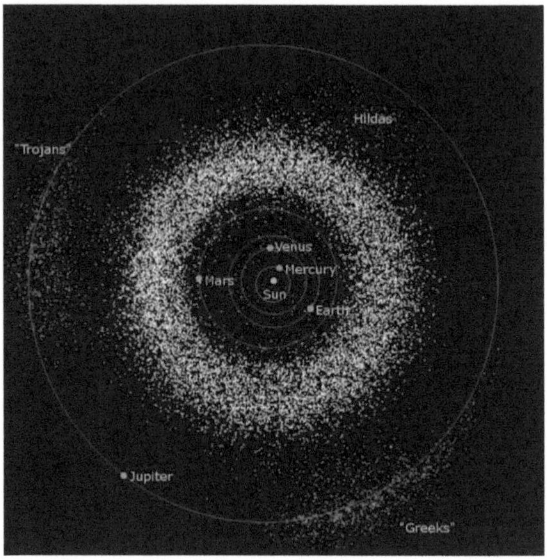

Trojan asteroids of Jupiter (coloured green) in front of and behind the planet along its orbital path. Also shown is the asteroid belt between the orbits of Mars and Jupiter (white), and the Hilda family of asteroids (brown).

Subsequently objects have been found orbiting the Lagrangian points of Neptune, Mars, and Earth.[5] Asteroids at the Lagrangian points of planets other than Jupiter may be called Lagrangian asteroids.[6]

- 5261 Eureka, 1998 VF_{31}, 1999 UJ_7, and 2007 NS_2 are Mars trojans.[7]
- Eight Neptune trojans[8] are known, but they may outnumber the Jupiter Trojans by an order of magnitude.[9] [10]
- 2010 TK_7 was confirmed as the first known Earth trojan in 2011. It is located in the L_4 Lagrangian point, which lies ahead of Earth.[11]

See also

- Lissajous orbit
- List of objects at Lagrangian points

References

[1] Robutel, P.; Souchay, J. (2010), "An introduction to the dynamics of trojan asteroids" (http://books.google.com/books?id=CLUYgQlWz4IC&pg=PA197), in Dvorak, Rudolf; Souchay, Jean, *Dynamics of Small Solar System Bodies and Exoplanets*, Lecture Notes in Physics, **790**, Springer, p. 197, ISBN 3642044573,

[2] NASA's WISE Mission Finds First Trojan Asteroid Sharing Earth's Orbit 7.27.11 (http://www.nasa.gov/mission_pages/WISE/news/wise20110727.html)

[3] Wright, Alison (August 1, 2011). "Planetary science: The Trojan is out there" (http://www.nature.com/nphys/journal/v7/n8/full/nphys2061.html). *Nature Physics* **7** (8): 592. Bibcode 2011NatPh...7..592W. doi:10.1038/nphys2061. . Retrieved 2011-08-12.

[4] Yoshida, F.; Nakamura, T (2005). "Size distribution of faint L_4 Trojan asteroids". *The Astronomical journal* **130** (6): 2900–11. Bibcode 2005AJ....130.2900Y. doi:10.1086/497571.

[5] Connors, Martin; Wieger, Paul; Veillet, Christian (27 July 2011). "Earth's Trojan asteroid" (http://www.nature.com/nature/journal/v475/n7357/full/nature10233.html). *Nature* **475** (7357): 481–483. Bibcode 2011Natur.475..481C. doi:10.1038/nature10233. . Retrieved

2011-07-27.
[6] Robert J. Whiteley and David J. Tholen, "A CCD Search for Lagrangian Asteroids of the Earth–Sun System", *Icarus* 136:1, November 1998:154–167
[7] "List of Martian Trojans" (http://www.minorplanetcenter.org/iau/lists/MarsTrojans.html). . Retrieved 2010-10-27.
[8] "List of Neptune Trojans" (http://www.minorplanetcenter.org/iau/lists/NeptuneTrojans.html). . Retrieved 2010-10-27.
[9] Chiang, E. I. & Lithwick, Y. *Neptune Trojans as a Testbed for Planet Formation*, The Astrophysical Journal, **628**, pp. 520–532 Preprint (http://www.arxiv.org/abs/astro-ph/0502276)
[10] David Powell (30 January 2007). "Neptune May Have Thousands of Escorts" (http://www.space.com/scienceastronomy/070130_st_neptune_trojans.html). Space.com. . Retrieved 2007-03-08.
[11] Choi, Charles Q. (27 July 2011). "First Asteroid Companion of Earth Discovered at Last" (http://www.space.com/12443-earth-asteroid-companion-discovered-2010-tk7.html). Space.com. . Retrieved 2011-07-27.

Minor_planet

A **minor planet** is an astronomical object in direct orbit around the Sun that is neither a dominant planet nor a comet, and thus includes the dwarf planets. Most minor planets are found in particular orbital niches; namely the asteroids, trojans, centaurs, Kuiper belt objects, and other trans-Neptunian objects. The first minor planet discovered was Ceres in 1801. The orbits of more than 540,000 objects have been archived at the Minor Planet Center.[1]

The term "minor planet" has been used since the 19th century to describe these objects.[2] The term **planetoid** has also been used, especially for larger objects.[3] Historically, the terms *asteroid*, *minor planet*, and *planetoid* have been more or less synonymous,[3] [4] but the issue has been complicated by the discovery of numerous minor planets beyond the orbit of Jupiter and especially Neptune that are not universally considered asteroids.[4]

Before 2006 the International Astronomical Union had officially used the term *minor planet*. During its 2006 meeting, the Union reclassified minor planets and comets into dwarf planets and small Solar System bodies.[5] Objects are called dwarf planets if their self-gravity is sufficient to achieve hydrostatic equilibrium, that is, an ellipsoidal shape, with all other minor planets and comets called "small Solar System bodies".[5] The IAU states: "the term 'minor planet' may still be used, but generally the term 'Small Solar System body' will be preferred."[6] However, for purposes of numbering and naming, the traditional distinction between minor planet and comet is still followed.

The Saturnian moon Mimas is the smallest body known to be in hydrostatic equilibrium (though not eligible to be a dwarf planet since it does not directly orbit the Sun), while the asteroid Pallas may be the largest that is not. The IAU has so far officially classified five objects as dwarf planets. In order of distance from the Sun, they are Ceres, Pluto, Haumea, Makemake, and Eris.

Populations

Hundreds of thousands of minor planets have been discovered within the Solar System, with the 2009 rate of discovery at over 3,000 per month. Of the more than 535,000 registered minor planets, 251,651 have orbits known well enough to be assigned permanent official numbers.[7] [1] Of these, 16,154 have official names[1] . As of September 2010, the lowest-numbered unnamed minor planet is (3708) 1974 FV_1;[8] but there are also some named minor planets above number 240,000.[9]

There are various broad minor-planet populations:

- Asteroids; traditionally, most have been bodies in the inner Solar System.[4]
  - Main-belt asteroids, those following roughly circular orbits between Mars and Jupiter. These are the original and best-known group of asteroids or minor planets.
  - Near-Earth asteroids, those whose orbits take them inside the orbit of Mars. Further subclassification of these, based on orbital distance, is traditional:

        - Aten asteroids, those that have semi-major axes of less than one Earth orbit. Those Aten asteroids that have their aphelion within Earth's orbit are known as Apohele asteroids;
        - Amor asteroids are those near-Earth asteroids that approach the orbit of the Earth from beyond, but do not cross it. Amor asteroids are further subdivided into four subgroups, depending on where their semimajor axis falls between Earth's orbit and the asteroid belt;
        - Apollo asteroids are those asteroids with a semimajor axis greater than Earth's, while having a perihelion distance of 1.017 AU or less. Like Aten asteroids, Apollo asteroids are Earth-crossers.
    - Earth trojans, asteroids sharing Earth's orbit and gravitationally locked to it. The only one known so far is 2010 TK_7.
    - Mars trojans, asteroids sharing Mars's orbit and gravitationally locked to it. Only four such asteroids are known.
    - Jupiter trojans, asteroids sharing Jupiter's orbit and gravitationally locked to it. Numerically they are estimated to equal the main-belt asteroids.
- Distant minor planets; an umbrella term for minor planets in the outer Solar System.
    - Centaurs, bodies in the outer Solar System between Jupiter and Neptune. They have unstable orbits due to the gravitational influence of the giant planets, and therefore must have come from elsewhere, probably outside Neptune.[10]
    - Neptune trojans, bodies sharing Neptune's orbit and gravitationally locked to it. Although only a handful are known, there is evidence that Neptune trojans are more numerous than either the asteroids in the asteroid belt or the Jupiter trojans.[11]
    - Trans-Neptunian objects, bodies at or beyond the orbit of Neptune, the outermost planet.
        - The Kuiper belt, objects inside an apparent population drop-off approximately 55 AU from the Sun.
            - Classical Kuiper belt objects, also known as cubewanos, are in primordial, relatively circular orbits that are not in resonance with Neptune.
            - Plutinos, bodies like Pluto that are in a 2:3 resonance with Neptune, and other resonant trans-Neptunian objects. This is a rather large group.
        - Scattered disc, objects with aphelia outside the Kuiper belt. These are thought to have been scattered by Neptune.
        - Detached objects such as Sedna, with both aphelia and perihelia outside the Kuiper belt.
        - The Oort Cloud, a hypothetical population thought to be the source of long-period comets that may extend out to 50,000 AU from the Sun.

Naming

A newly discovered minor planet is given a provisional designation (such as 2002 AT_4) consisting of the year of discovery and an alphanumeric code indicating the half-month of discovery and the sequence within that half-month. Once an asteroid's orbit has been confirmed, it is given a number, and later may also be given a name (e.g. 433 Eros). The formal naming convention uses parentheses around the number, but dropping the parentheses is quite common. Informally, it is common to drop the number altogether, or to drop it after the first mention when a name is repeated in running text.

Minor planets that have been given a number but not a name keep their provisional designation, e.g. (29075) 1950 DA. As modern discovery techniques are finding vast numbers of new asteroids, they are increasingly being left unnamed. The earliest discovered to be left unnamed was for a long time (3360) 1981 VA, now 3360 Syrinx; as of September 2008, this distinction is held by (3708) 1974 FV_1. On rare occasions, a small object's provisional designation may become used as a name in itself: the still unnamed (15760) 1992 QB_1 gave its "name" to a group of Kuiper belt objects which became known as cubewanos.

A few objects are cross-listed as both comets and asteroids, such as 4015 Wilson–Harrington, which is also listed as 107P/Wilson–Harrington.

Numbering

Minor planets are awarded an official number once their orbits are confirmed. With the increasing rapidity of discovery, these are now six-figure numbers. The switch from five figures to six figures arrived with the publication of the Minor Planet Circular (MPC) of October 19, 2005, which saw the highest numbered minor planet jump from 99947 to 118161.

Sources for names

The first few asteroids were named after figures from Greek and Roman mythology, but as such names started to dwindle the names of famous people, literary characters, discoverer's wives, children, and even television characters were used.

The first asteroid to be given a non-mythological name was 20 Massalia, named after the Greek name for the city of Marseilles. The first to be given an entirely non-Classical name was 45 Eugenia, named after Empress Eugénie de Montijo, the wife of Napoleon III. For some time only female (or feminized) names were used; Alexander von Humboldt was the first man to have an asteroid named after him, but his name was feminized to 54 Alexandra. This unspoken tradition lasted until 334 Chicago was named; even then, oddly feminised names show up in the list for years after.

As the number of asteroids began to run into the hundreds, and eventually the thousands, discoverers began to give them increasingly frivolous names. The first hints of this were 482 Petrina and 483 Seppina, named after the discoverer's pet dogs. However, there was little controversy about this until 1971, upon the naming of 2309 Mr. Spock (the name of the discoverer's cat). Although the IAU subsequently banned pet names as sources, eccentric asteroid names are still being proposed and accepted, such as 4321 Zero, 6042 Cheshirecat, 9007 James Bond, 13579 Allodd and 24680 Alleven, and 26858 Misterrogers.

A well-established rule is that, unlike comets, minor planets may not be named after their discoverer(s). One way to circumvent this rule has been for astronomers to exchange the courtesy of naming their discoveries after each other. An exception to this rule is 96747 Crespodasilva, which was named after its discoverer, Lucy d'Escoffier Crespo da Silva, because she died shortly after the discovery, at age 22.[12] [13]

Names were adapted to various languages from the beginning. 1 Ceres, *Ceres* being its Anglo-Latin name, was actually named *Cerere*, the Italian form of the name. German, French, Arabic and Hindi use forms similar to the English, whereas Russian uses a form, *Tserera*, similar to the Italian. In Greek the name was translated to Demeter, the Greek equivalent of the Roman goddess Ceres. In the early years, before it started causing conflicts, asteroids named after Roman figures were generally translated in Greek; other examples are Hera for 3 Juno, Hestia for 4 Vesta, Chloris for 8 Flora, and Pistê for 37 Fides. In Chinese, the names are not given the Chinese forms of the deities they are named after, but rather typically have a syllable or two for the character of the deity or person, followed by 'god(dess)' or 'woman' if just one syllable, plus 'star/planet', so that most asteroid names are written with three Chinese characters. Thus Ceres is 'grain goddess planet',[14] Pallas is 'wisdom goddess planet', etc.

Special naming rules

Minor-planet naming is not always a free-for-all: there are some populations for which rules have developed about the sources of names. For instance, centaurs (orbiting between Saturn and Neptune) are all named after mythological centaurs; Jupiter trojans after heroes from the Trojan War; resonant trans-Neptunian objects after underworld spirits; and non-resonant TNOs after creation deities.

Physical properties of comets and minor planets

Commission 15[15] of the International Astronomical Union is dedicated to the Physical Study of Comets & Minor Planets.

Archival data on the physical properties of comets and minor planets are found in the PDS Asteroid/Dust Archive.[16] This includes standard asteroid physical characteristics such as the properties of binary systems, occultation timings and diameters, masses, densities, rotation periods, surface temperatures, albedoes, spin vectors, taxonomy, and absolute magnitudes and slopes. In addition, European Asteroid Research Node (E.A.R.N.), an association of asteroid research groups, maintains a Data Base of Physical and Dynamical Properties of Near Earth Asteroids.[17]

Most detailed information is available from Category: Asteroids visited by spacecraft and Category: Comets visited by spacecraft.

See also

- List of minor planets
- Solar System
- Groups of minor planets
- Small Solar System body
- Plutoid

External links

- Minor Planet Center [18]

References

[1] "Minor Planet Statistics" (http://www.minorplanetcenter.net/iau/lists/ArchiveStatistics.html). Minor Planet Center. . Retrieved 2011-02-26.

[2] When did the asteroids become minor planets? (http://aa.usno.navy.mil/faq/docs/minorplanets.php), James L. Hilton, Astronomical Information Center, United States Naval Observatory. Accessed May 5, 2008.

[3] Planet, asteroid, minor planet: A case study in astronomical nomenclature, David W. Hughes, Brian G. Marsden, *Journal of Astronomical History and Heritage* **10**, #1 (2007), pp. 21–30. Bibcode: 2007JAHH...10...21H

[4] " Asteroid (http://encarta.msn.com/encyclopedia_761551567/asteroid.html)", *MSN Encarta*, Microsoft. Accessed May 5, 2008. Archived (http://www.webcitation.org/5kx3nDxZi) 2009-11-01.

[5] Press release, IAU 2006 General Assembly: Result of the IAU Resolution votes (http://www.astronomy2006.com/press-release-24-8-2006-2.php), International Astronomical Union, August 24, 2006. Accessed May 5, 2008.

[6] Questions and Answers on Planets (http://www.iau.org/public_press/news/release/iau0603/questions_answers/), additional information, news release IAU0603, IAU 2006 General Assembly: Result of the IAU Resolution votes, International Astronomical Union, August 24, 2006. Accessed May 8, 2008.

[7] JPL. "How Many Solar System Bodies" (http://ssd.jpl.nasa.gov/?body_count). *JPL Solar System Dynamics*. NASA. . Retrieved 2010-09-27.

[8] "Discovery Circumstances: Numbered Minor Planets (1)-(5000)" (http://www.minorplanetcenter.net/iau/lists/NumberedMPs000001.html). Minor Planet Center. . Retrieved 2011-02-26.

[9] "Discovery Circumstances: Numbered Minor Planets (240001)-(245000)" (http://www.minorplanetcenter.net/iau/lists/NumberedMPs240001.html). Minor Planet Center. . Retrieved 2011-02-26.

[10] Horner, J.; Evans, N.W.; Bailey, M. E. (2004). "Simulations of the Population of Centaurs I: The Bulk Statistics". *Monthly Notices of the Royal Astronomical Society* **354** (3): 798–810. arXiv:astro-ph/0407400. Bibcode 2004MNRAS.354..798H. doi:10.1111/j.1365-2966.2004.08240.x.

[11] Neptune trojans, Jupiter trojans (http://www.dtm.ciw.edu/users/sheppard/trojans/)

[12] NASA JPL Small-Body Database Browser on 96747 Crespodasilva (http://ssd.jpl.nasa.gov/sbdb.cgi?sstr=96747+Crespodasilva)

[13] Staff (November 28, 2000). "Lucy Crespo da Silva, 22, a senior, dies in fall" (http://hubblesite.org/newscenter/newsdesk/archive/releases/1991/12/text/). Hubble News Desk. . Retrieved 2008-04-15.

[14] 'valley' being a common abbreviation of 'grain' that would be formally adopted with simplified Chinese characters.

[15] "Division III Commission 15 Physical Study of Comets & Minor Planets" (http://www.iau.org/science/scientific_bodies/commissions/15). International Astronomical Union (IAU). September 29, 2005. . Retrieved 2010-03-22.

[16] "Physical Properties of Asteroids" (http://www.psi.edu/pds/archive/physical.html). .

[17] "The Near-Earth Asteroids Data Base" (http://earn.dlr.de/nea). .

[18] http://www.minorplanetcenter.net/iau/mpc.html

rue:Планетка

Kuiper_belt

The **Kuiper belt** (/ˈkaɪpər/, rhyming with "viper"), sometimes called the **Edgeworth–Kuiper belt**, is a region of the Solar System beyond the planets extending from the orbit of Neptune (at 30 AU) to approximately 50 AU from the Sun.[1] It is similar to the asteroid belt, although it is far larger—20 times as wide and 20 to 200 times as massive.[2] [3] Like the asteroid belt, it consists mainly of small bodies, or remnants from the Solar System's formation. While the asteroid belt is composed primarily of rock, ices, and metal, the Kuiper objects are composed largely of frozen volatiles (termed "ices"), such as methane, ammonia and water. The classical (low-eccentricity) belt is home to at least three dwarf planets: Pluto, Haumea, and Makemake. Some of the Solar System's moons, such as Neptune's Triton and Saturn's Phoebe, are also believed to have originated in the region.[4] [5]

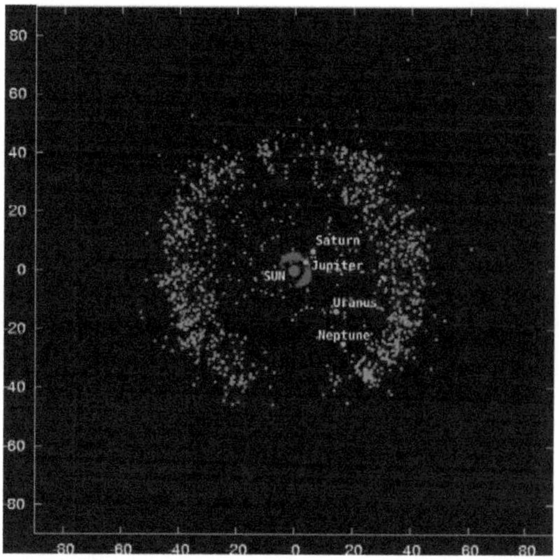

Known objects in the Kuiper belt, derived from data from the Minor Planet Center. Objects in the main belt are coloured green, while scattered objects are coloured orange. The four outer planets are blue. Neptune's few known trojans are yellow, while Jupiter's are pink. The scattered objects between Jupiter's orbit and the Kuiper belt are known as centaurs. The scale is in astronomical units. The pronounced gap at the bottom is due to difficulties in detection against the background of the plane of the Milky Way.

Since the belt was discovered in 1992,[6] the number of known **Kuiper belt objects** (**KBOs**) has increased to over a thousand, and more than 70,000 KBOs over 100 km (62 mi) in diameter are believed to exist.[7] The Kuiper belt was initially believed to be the main repository for periodic comets, those with orbits lasting less than 200 years. However, studies since the mid-1990s have shown that the classical belt is dynamically stable, and that comets' true place of origin is the scattered disc, a dynamically active region created by the outward motion of Neptune 4.5 billion years ago;[8] scattered disc objects such as Eris have extremely eccentric orbits that take them as far as 100 AU from the Sun.[9]

Pluto is the largest known member of the Kuiper belt, if the scattered disc is excluded. Originally considered a planet, Pluto's position as part of the Kuiper belt has caused it to be reclassified as a "dwarf planet". It is compositionally similar to many other objects of the Kuiper belt, and its orbital period is identical to that of the KBOs known as "plutinos". In Pluto's honour, the four currently accepted dwarf planets beyond Neptune's orbit are called "plutoids".

The Kuiper belt should not be confused with the hypothesized Oort cloud, which is a thousand times more distant. The objects within the Kuiper belt, together with the members of the scattered disc and any potential Hills cloud or Oort cloud objects, are collectively referred to as trans-Neptunian objects (TNOs).[10]

History

Since the discovery of Pluto, many have speculated that it might not be alone. The region now called the Kuiper belt had been hypothesized in various forms for decades. It was only in 1992 that the first direct evidence for its existence was found. The number and variety of prior speculations on the nature of the Kuiper belt have led to continued uncertainty as to who deserves credit for first proposing it.

Hypotheses

The first astronomer to suggest the existence of a trans-Neptunian population was Frederick C. Leonard. In 1930, soon after Pluto's discovery by Clyde Tombaugh, Leonard pondered whether it was "not likely that in Pluto there has come to light the *first* of a *series* of ultra-Neptunian bodies, the remaining members of which still await discovery but which are destined eventually to be detected".[11]

In 1943, in the *Journal of the British Astronomical Association*, Kenneth Edgeworth hypothesized that, in the region beyond Neptune, the material within the primordial solar nebula was too widely spaced to condense into planets, and so rather condensed into a myriad of smaller bodies. From this he concluded that "the outer region of the solar system, beyond the orbits of the planets, is occupied by a very large number of comparatively small bodies"[12] and that, from time to time, one of their number "wanders from its own sphere and appears as an occasional visitor to the inner solar system",[13] becoming a comet.

Astronomer Gerard Kuiper, after whom the Kuiper belt is named

In 1951, in an article for the journal *Astrophysics*, Gerard Kuiper speculated on a similar disc having formed early in the Solar System's evolution; however, he did not believe that such a belt still existed today. Kuiper was operating on the assumption common in his time, that Pluto was the size of the Earth, and had therefore scattered these bodies out toward the Oort cloud or out of the Solar System. Were Kuiper's hypothesis correct, there would not be a Kuiper belt where we now see it.[14]

The hypothesis took many other forms in the following decades: in 1962, physicist Al G.W. Cameron postulated the existence of "a tremendous mass of small material on the outskirts of the solar system",[15] while in 1964, Fred Whipple, who popularised the famous "dirty snowball" hypothesis for cometary structure, thought that a "comet belt" might be massive enough to cause the purported discrepancies in the orbit of Uranus that had sparked the search for Planet X, or at the very least, to affect the orbits of known comets.[16] Observation, however, ruled out this hypothesis.[15]

In 1977, Charles Kowal discovered 2060 Chiron, an icy planetoid with an orbit between Saturn and Uranus. He used a blink comparator; the same device that had allowed Clyde Tombaugh to discover Pluto nearly 50 years before.[17] In 1992, another object, 5145 Pholus, was discovered in a similar orbit.[18] Today, an entire population of comet-like bodies, the centaurs, is known to exist in the region between Jupiter and Neptune. The centaurs' orbits are unstable and have dynamical lifetimes of a few million years.[19] From the time of Chiron's discovery, astronomers speculated that they therefore must be frequently replenished by some outer reservoir.[20]

Further evidence for the belt's existence later emerged from the study of comets. That comets have finite lifespans has been known for some time. As they approach the Sun, its heat causes their volatile surfaces to sublimate into space, eating them gradually away. In order to still be visible over the age of the Solar System, they must be frequently replenished.[21] One such area of replenishment is the Oort cloud, the spherical swarm of comets extending beyond 50 000 AU from the Sun first hypothesised by astronomer Jan Oort in 1950.[22] It is believed to be the point of origin for long period comets, those, like Hale-Bopp, with orbits lasting thousands of years.

There is however another comet population, known as short period or periodic comets; those, like Halley, with orbits lasting less than 200 years. By the 1970s, the rate at which short-period comets were being discovered was becoming increasingly inconsistent with them having emerged solely from the Oort cloud.[23] For an Oort cloud object to become a short-period comet, it would first have to be captured by the giant planets. In 1980, in the Monthly Notices of the Royal Astronomical Society, Julio Fernandez stated that for every short period comet to be sent into the inner solar system from the Oort cloud, 600 would have to be ejected into interstellar space. He speculated that a comet belt from between 35 and 50 AU would be required to account for the observed number of comets.[24] Following up on Fernandez's work, in 1988 the Canadian team of Martin Duncan, Tom Quinn and Scott Tremaine ran a number of computer simulations to determine if all observed comets could have arrived from the Oort cloud. They found that the Oort cloud could not account for all short-period comets, particularly as short-period comets are clustered near the plane of the Solar System, whereas Oort cloud comets tend to arrive from any point in the sky. With a belt as Fernandez described it added to the formulations, the simulations matched observations.[25] Reportedly because the words "Kuiper" and "comet belt" appeared in the opening sentence of Fernandez's paper, Tremaine named this hypothetical region the "Kuiper belt".[26]

Discovery

The array of telescopes atop Mauna Kea, with which the Kuiper belt was discovered

In 1987, astronomer David Jewitt, then at MIT, became increasingly puzzled by "the apparent emptiness of the outer Solar System".[6] He encouraged then-graduate student Jane Luu to aid him in his endeavour to locate another object beyond Pluto's orbit, because, as he told her, "If we don't, nobody will.".[27] Using telescopes at the Kitt Peak National Observatory in Arizona and the Cerro Tololo Inter-American Observatory in Chile, Jewitt and Luu conducted their search in much the same way as Clyde Tombaugh and Charles Kowal had, with a blink comparator.[27] Initially, examination of each pair of plates took about eight hours,[28] but the process was sped up with the arrival of electronic charge-coupled devices or CCDs, which, though their field of view was narrower, were not only more efficient at collecting light (they retained 90 percent of the light that hit them, rather than the ten percent achieved by photographs) but allowed the blinking process to be done virtually, on a computer screen. Today, CCDs form the basis for most astronomical detectors.[29] In 1988, Jewitt moved to the Institute of Astronomy at the University of Hawaii. Luu later joined him to work at the University of Hawaii's 2.24 m telescope at Mauna Kea.[30] Eventually, the field of view for CCDs had increased to 1024 by 1024 pixels, which allowed searches to be conducted far more rapidly.[31] Finally, after five years of searching, on August 30, 1992, Jewitt and Luu announced the "Discovery of the candidate Kuiper belt object" (15760) 1992 QB_1.[6] Six months later, they discovered a second object in the region, (181708) 1993 FW.[32]

Studies since the trans-Neptunian region was first charted have shown that in fact, the region now called the Kuiper belt is not the point of origin for short-period comets, but that they instead derive from a linked population called the scattered disc. The scattered disc was created when Neptune migrated outward into the proto-Kuiper belt, which at the time was much closer to the Sun, and left in its wake a population of dynamically stable objects which could never be affected by its orbit (the Kuiper belt proper), and a population whose perihelia are close enough that Neptune can still disturb them as it travels around the Sun (the scattered disc). Because the scattered disc is dynamically active and the Kuiper belt relatively dynamically stable, the scattered disc is now seen as the most likely point of origin for periodic comets.[8]

Name

Astronomers will sometimes use alternative name **Edgeworth–Kuiper belt** to credit Edgeworth, and KBOs are occasionally referred to as EKOs. However, Brian Marsden claims neither deserve true credit; "Neither Edgeworth or Kuiper wrote about anything remotely like what we are now seeing, but Fred Whipple did."[33] Conversely, David Jewitt comments that, "If anything ... Fernandez most nearly deserves the credit for predicting the Kuiper Belt."[14] The term **trans-Neptunian object** (TNO) is recommended for objects in the belt by several scientific groups because the term is less controversial than all others—it is not a synonym though, as TNOs include all objects orbiting the Sun past the orbit of Neptune, not just those in the Kuiper belt.

Origins

The precise origins of the Kuiper belt and its complex structure are still unclear, and astronomers are awaiting the completion of several wide-field survey telescopes such as Pan-STARRS and the future LSST, which should reveal many currently unknown KBOs. These surveys will provide data that will help determine answers to these questions.[2]

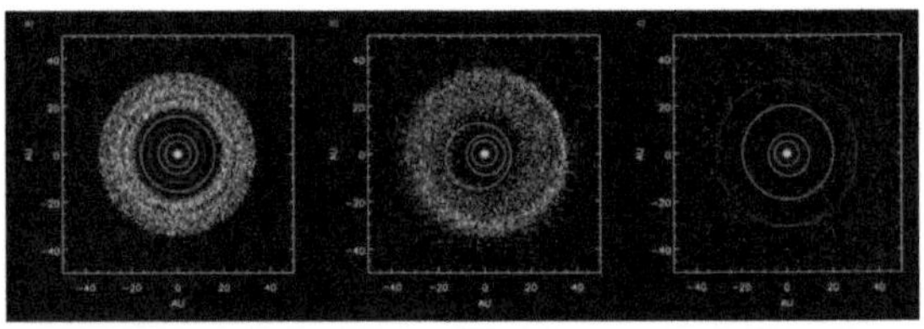

Simulation showing outer planets and Kuiper belt: a) before Jupiter/Saturn 2:1 resonance, b) scattering of Kuiper belt objects into the solar system after the orbital shift of Neptune, c) after ejection of Kuiper Belt bodies by Jupiter

The Kuiper belt is believed to consist of planetesimals; fragments from the original protoplanetary disc around the Sun that failed to fully coalesce into planets and instead formed into smaller bodies, the largest less than 3000 kilometres (1900 mi) in diameter.

Modern computer simulations show the Kuiper belt to have been strongly influenced by Jupiter and Neptune, and also suggest that neither Uranus nor Neptune could have formed *in situ* beyond Saturn, as too little primordial matter existed at that range to produce objects of such high mass. Instead, these planets are believed to have formed closer to Jupiter, and migrated outwards during the course of the Solar System's early evolution. Eventually, the orbits shifted to the point where Jupiter and Saturn existed in an exact 2:1 resonance; Jupiter orbited the Sun twice for every one Saturn orbit. The gravitational pull from such a resonance ultimately disrupted the orbits of Uranus and Neptune, causing Neptune's orbit to move outward into the primordial planetesimal disk, which sent the disk into temporary chaos.[34] As Neptune traveled along this modified orbit, it excited and scattered many TNO planetesimals into higher and more eccentric orbits, depleting the primordial population.[35]

However, the present most popular model still fails to account for many of the characteristics of the distribution and, quoting one of the scientific articles,[36] the problems "continue to challenge analytical techniques and the fastest numerical modeling hardware and software".

The frequency of paired objects, many of which are far apart and loosely bound, also poses a problem for the Nice model of formation.[37]

Structure

At its fullest extent, including its outlying regions, the Kuiper belt stretches from roughly 30 to 55 AU. However, the main body of the belt is generally accepted to extend from the 2:3 resonance (see below) at 39.5 AU to the 1:2 resonance at roughly 48 AU.[38] The Kuiper belt is quite thick, with the main concentration extending as much as ten degrees outside the ecliptic plane and a more diffuse distribution of objects extending several times farther. Overall it more resembles a torus or doughnut than a belt.[39] Its mean position is inclined to the ecliptic by 1.86 degrees.[40]

Dust in the Kuiper belt creates a faint infrared disk.

The presence of Neptune has a profound effect on the Kuiper belt's structure due to orbital resonances. Over a timescale comparable to the age of the Solar System, Neptune's gravity destabilises the orbits of any objects which happen to lie in certain regions, and either sends them into the inner Solar System or out into the scattered disc or interstellar space. This causes the Kuiper belt to possess pronounced gaps in its current layout, similar to the Kirkwood gaps in the asteroid belt. In the region between 40 and 42 AU, for instance, no objects can retain a stable orbit over such times, and any observed in that region must have migrated there relatively recently.[41]

Classical belt

Between the 2:3 and 1:2 resonances with Neptune, at approximately 42–48 AU, the gravitational influence of Neptune is negligible, and objects can exist with their orbits essentially unmolested. This region is known as the classical Kuiper belt, and its members comprise roughly two thirds of KBOs observed to date.[42] [43] Because the first modern KBO discovered, 1992 QB1, is considered the prototype of this group, classical KBOs are often referred to as cubewanos ("Q-B-1-os").[44] [45] The guidelines established by the IAU demand that classical KBOs be given names of mythological beings associated with creation.[46]

The classical Kuiper belt appears to be a composite of two separate populations. The first, known as the "dynamically cold" population, has orbits much like the planets; nearly circular, with an orbital eccentricity of less than 0.1, and with relatively low inclinations up to about 10° (they lie close to the plane of the Solar System rather than at an angle). The second, the "dynamically hot" population, has orbits much more inclined to the ecliptic, by up to 30°. The two populations have been named this way not because of any major difference in temperature, but from analogy to particles in a gas, which increase their relative velocity as they become heated up.[47] The two populations not only possess different orbits, but different compositions; the cold population is markedly redder than the hot, suggesting it formed in a different region. The hot population is believed to have formed near Jupiter, and to have been ejected out by movements among the gas giants. The cold population, on the other hand, is believed to have formed more or less in its current position although it may also have been later swept outwards by Neptune during its migration.[2] [48]

Resonances

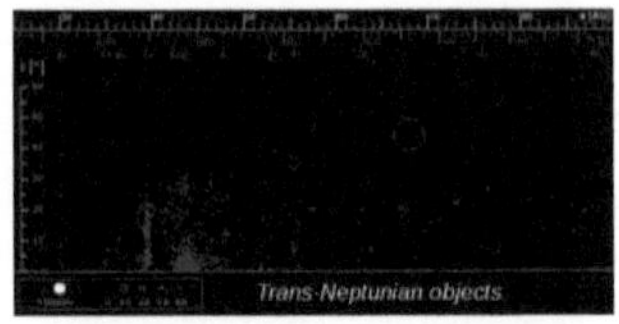

Distribution of cubewanos (blue), Resonant trans-Neptunian objects (red) and near scattered objects (grey).

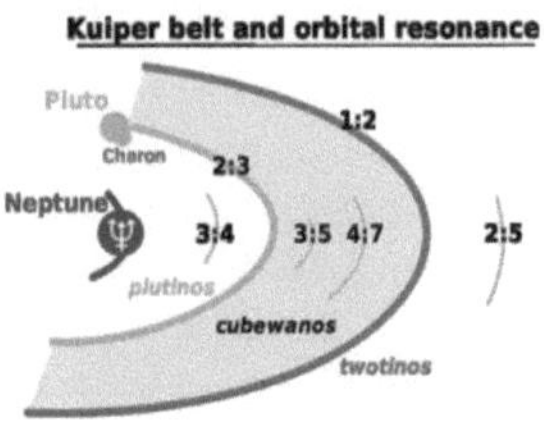

Orbit classification (schematic of semi-major axes).

When an object's orbital period is an exact ratio of Neptune's (a situation called a mean motion resonance), then it can become locked in a synchronised motion with Neptune and avoid being perturbed away if their relative alignments are appropriate. If, for instance, an object is in just the right kind of orbit so that it orbits the Sun two times for every three Neptune orbits, and if it reaches perihelion with Neptune a quarter of an orbit away from it, then whenever it returns to perihelion, Neptune will always be in about the same relative position as it began, since it will have completed 1½ orbits in the same time. This is known as the 2:3 (or 3:2) resonance, and it corresponds to a characteristic semi-major axis of about 39.4 AU. This 2:3 resonance is populated by about 200 known objects,[49] including Pluto together with its moons. In recognition of this, the other members of this family are known as plutinos. Many plutinos, including Pluto, often have orbits which cross that of Neptune, though their resonance means they can never collide. Many others, such as 90482 Orcus and 28978 Ixion, are large enough to probably qualify as plutoids when more is known about them.[50] [51] Plutinos have high orbital eccentricities, suggesting that they are not native to their current positions but were instead thrown haphazardly into their orbits by the migrating Neptune.[52] IAU guidelines dictate that all plutinos must, like Pluto, be named for underworld deities.[46] The 1:2 resonance (whose objects complete half an orbit for each of Neptune's) corresponds to semi-major axes of ~47.7AU, and is sparsely populated.[53] Its residents are sometimes referred to as twotinos. Other resonances also exist at 3:4, 3:5, 4:7 and 2:5.[54] Neptune possesses a number of trojan objects, which occupy its L_4 and L_5 points; gravitationally stable regions leading and trailing it in its orbit. Neptune trojans are often described as being in a 1:1 resonance with Neptune. Neptune trojans are remarkably stable in their orbits and are unlikely to have been captured by Neptune, but rather to have formed alongside it.[52]

Additionally, there is a relative absence of objects with semi-major axes below 39 AU which cannot apparently be explained by the present resonances. The currently accepted hypothesis for the cause of this is that as Neptune migrated outward, unstable orbital resonances moved gradually through this region, and thus any objects within it were swept up, or gravitationally ejected from it.[55]

"Kuiper cliff"

The 1:2 resonance appears to be an edge beyond which few objects are known. It is not clear whether it is actually the outer edge of the classical belt or just the beginning of a broad gap. Objects have been detected at the 2:5 resonance at roughly 55 AU, well outside the classical belt; however, predictions of a large number of bodies in classical orbits between these resonances have not been verified through observation.[52]

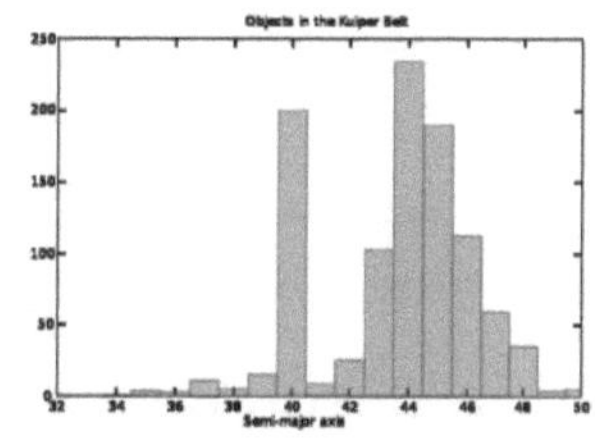

Graph showing the numbers of KBOs for a given distance from the Sun. The plutinos are the "spike" at 40 AU, while the classicals are between 42 and 47 AU, and the twotinos are at 48 AU.

Earlier models of the Kuiper belt had suggested that the number of large objects would increase by a factor of two beyond 50 AU,[56] so this sudden drastic falloff, known as the "Kuiper cliff", was completely unexpected, and its cause, to date, is unknown. Bernstein and Trilling et al. have found evidence that the rapid decline in objects of 100 km or more in radius beyond 50 AU is real, and not due to observational bias. Possible explanations include that material at that distance is too scarce or too scattered to accrete into large objects, or that subsequent processes removed or destroyed those which did form.[57] Patryk Lykawka of Kobe University has claimed that the gravitational attraction of an unseen large planetary object, perhaps the size of Earth or Mars, might be responsible.[58] [59]

Composition

Studies of the Kuiper belt since its discovery have generally indicated that its members are primarily composed of ices: a mixture of light hydrocarbons (such as methane), ammonia, and water ice,[60] a composition they share with comets.[61] The low densities observed in those KBOs whose diameter is known, (less than 1 g cm^{-3}) is consistent with an icy makeup.[60] The temperature of the belt is only about 50K,[62] so many compounds that would be gaseous closer to the Sun remain solid.

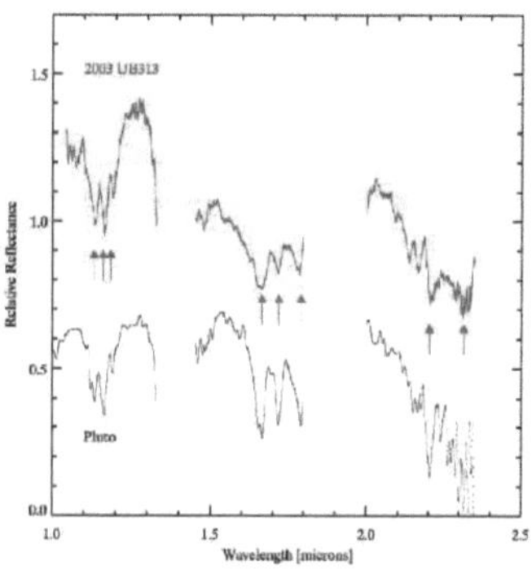

The infrared spectra of both Eris and Pluto, highlighting their common methane absorption lines

Due to their small size and extreme distance from Earth, the chemical makeup of KBOs is very difficult to determine. The principal method by which astronomers determine the composition of a celestial object is spectroscopy. When an object's light is broken into its component colours, an image akin to a rainbow is formed. This image is called a spectrum. Different substances absorb light at different wavelengths, and when the spectrum for a specific object is unravelled, dark lines (called absorption lines) appear where the substances within it have absorbed that particular wavelength of light. Every element or compound has its own unique spectroscopic signature, and by reading an object's full spectral "fingerprint", astronomers can determine what it is made of.

Initially, such detailed analysis of KBOs was impossible, and so astronomers were only able to determine the most basic facts about their makeup, primarily their colour.[63] These first data showed a broad range of colours among KBOs, ranging from neutral grey to deep red.[64] This suggested that their surfaces were composed of a wide range of compounds, from dirty ices to hydrocarbons.[64] This diversity was startling, as astronomers had expected KBOs to be uniformly dark, having lost most of their volatile ices to the effects of cosmic rays.[65] Various solutions were suggested for this discrepancy, including resurfacing by impacts or outgassing.[63] However, Jewitt and Luu's spectral analysis of the known Kuiper belt objects in 2001 found that the variation in colour was too extreme to be easily explained by random impacts.[66]

Although to date most KBOs still appear spectrally featureless due to their faintness, there have been a number of successes in determining their composition.[62] In 1996, Robert H. Brown *et al.* obtained spectroscopic data on the KBO 1993 SC, revealing its surface composition to be markedly similar to that of Pluto, as well as Neptune's moon Triton, possessing large amounts of methane ice.[67]

Water ice has been detected in several KBOs, including 1996 TO66,[68] 2000 EB173 and 2000 WR106.[69] In 2004, Mike Brown *et al.* determined the existence of crystalline water ice and ammonia hydrate on one of the largest known KBOs, 50000 Quaoar. Both of these substances would have been destroyed over the age of the solar system, suggesting that Quaoar had been recently resurfaced, either by internal tectonic activity or by meteorite impacts.[62]

Mass and size distribution

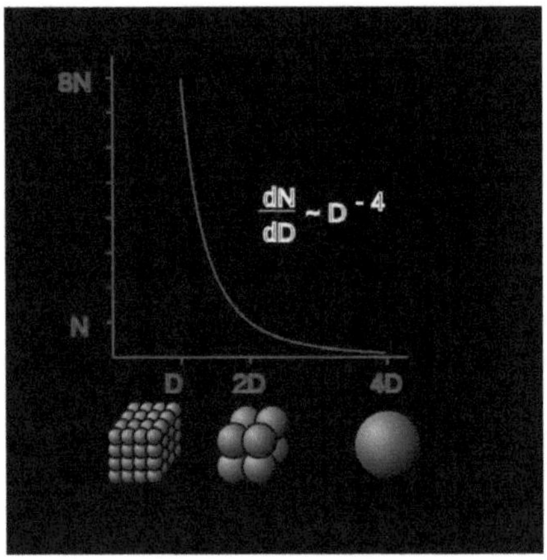

Illustration of the power law.

Despite its vast extent, the collective mass of the Kuiper belt is relatively low. The total mass is estimated at range between a 25th and 10th the mass of the Earth[70] with some estimates placing it at a thirtieth an Earth mass.[71] Conversely, models of the Solar System's formation predict a collective mass for the Kuiper belt of 30 Earth masses.[2] This missing >99% of the mass can hardly be dismissed, as it is required for the accretion of any KBOs larger than 100 km (62 mi) in diameter. If the Kuiper belt had always had its current low density these large objects simply could not have formed.[2] Moreover, the eccentricity and inclination of current orbits makes the encounters quite "violent," resulting in destruction rather than accretion. It appears that either the current residents of the Kuiper belt have been created closer to the Sun or some mechanism dispersed the original mass. Neptune's current influence is too weak to explain such a massive "vacuuming", though the Nice model proposes that it could have been the cause of mass removal in the past. While the question remains open, the conjectures vary from a passing star scenario to grinding of smaller objects, via collisions, into dust small enough to be affected by solar radiation.[48]

Bright objects are rare compared with the dominant dim population, as expected from accretion models of origin, given that only some objects of a given size would have grown further. This relationship N(D), the population expressed as a function of the diameter, referred to as brightness slope, has been confirmed by observations. The slope is inversely proportional to some power of the diameter D.

where the current measures[72] give q = 4 ±0.5.

Less formally, there are for instance 8 ($=2^3$) times more objects in 100–200 km range than objects in 200–400 km range. In other words, for every object with the diameter of 1000 km (621 mi) there should be around 1000 ($=10^3$) objects with diameter of 100 km (62 mi).

The law is expressed in this differential form rather than as a cumulative cubic relationship, because only the middle part of the slope can be measured; the law must break at smaller sizes, beyond the current measure.

Of course, only the magnitude is actually known, the size is inferred assuming albedo (not a safe assumption for larger objects).

Since January 2010, the smallest Kuiper belt object discovered to date spans 980 m across.[73]

Scattered objects

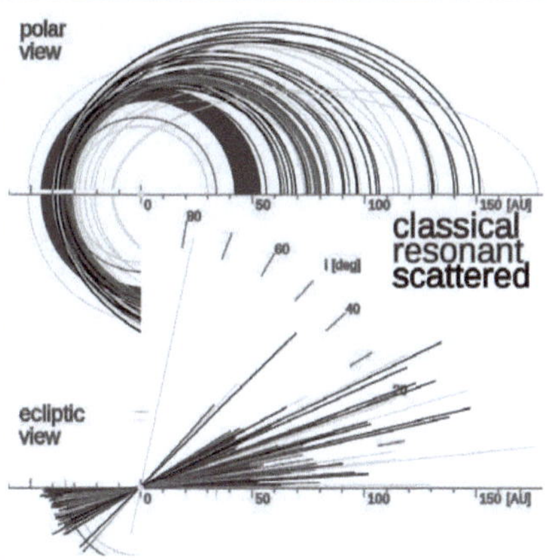

The orbits of objects in the scattered disc; the classical KBOs are blue, while the 2:5 resonant objects are green.

The scattered disc is a sparsely populated region, overlapping with the Kuiper belt but extending as far as 100 AU and farther. Scattered disc objects (SDOs) travel in highly elliptical orbits, usually also highly inclined to the ecliptic. Most models of solar system formation show both KBOs and SDOs first forming in a primordial comet belt, while later gravitational interactions, particularly with Neptune, sent the objects spiraling outward; some into stable orbits (the KBOs) and some into unstable orbits, becoming the scattered disc.[8] Due to its unstable nature, the scattered disc is believed to be the point of origin for many of the Solar System's short-period comets. Their dynamic orbits occasionally force them into the inner Solar System, becoming first centaurs, and then short-period comets.[8]

According to the Minor Planet Center, which officially catalogues all trans-Neptunian objects, a KBO, strictly speaking, is any object that orbits exclusively within the defined Kuiper belt region regardless of origin or composition. Objects found outside the belt are classed as scattered objects.[74] However, in some scientific circles the term "Kuiper belt object" has become synonymous with any icy planetoid native to the outer solar system believed to have been part of that initial class, even if its orbit during the bulk of solar system history has been beyond the Kuiper belt (e.g. in the scattered disc region). They often describe scattered disc objects as "scattered Kuiper belt objects."[75] Eris, the recently discovered object now known to be larger than Pluto, is often referred to as a KBO, but is technically an SDO.[74] A consensus among astronomers as to the precise definition of the Kuiper belt has yet to be reached, and this issue remains unresolved.

The centaurs, which are not normally considered part of the Kuiper belt, are also believed to be scattered objects, the only difference being that they were scattered inward, rather than outward. The Minor Planet Center groups the centaurs and the SDOs together as scattered objects.[74]

Triton

Neptune's moon Triton

During its period of migration, Neptune is thought to have captured one of the larger KBOs and set it in orbit around itself. This is its moon Triton, which is the only large moon in the Solar System to have a retrograde orbit; it orbits in the opposite direction to Neptune's rotation. This suggests that, unlike the large moons of Jupiter and Saturn, which are thought to have coalesced from spinning discs of material encircling their young parent planets, Triton was a fully formed body that was captured from surrounding space. Gravitational capture of an object is not easy; it requires that some force act upon the object to slow it down enough to be snared by the larger object's gravity. How this happened to Triton is not well understood, though it does suggest that Triton formed as part of a large population of similar objects whose gravity could impede its motion enough to be captured.[76] Triton is only slightly larger than Pluto, and spectral analysis of both worlds shows that they are largely composed of similar materials, such as methane and carbon monoxide. All this points to the conclusion that Triton was once a KBO that was captured by Neptune during its outward migration.[77]

Largest KBOs

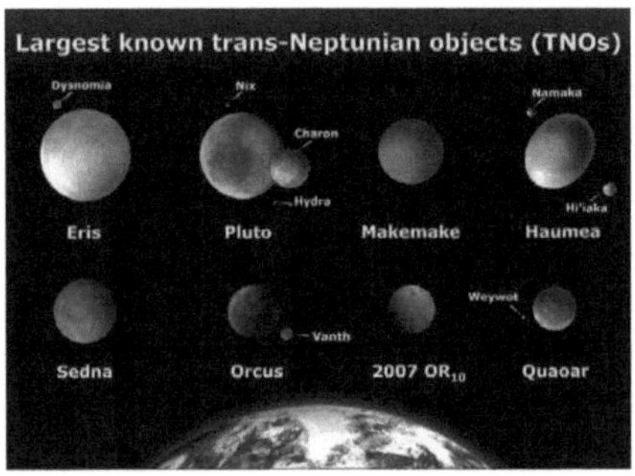

Artistic comparison of Eris, Pluto, Makemake, Haumea, Sedna, Orcus, 2007 OR10, Quaoar, and Earth (scales are outdated)

Since the year 2000, a number of KBOs with diameters of between 500 and 1500 km (932 mi), more than half that of Pluto, have been discovered. 50000 Quaoar, a classical KBO discovered in 2002, is over 1,200 km across. Makemake (originally (136472) 2005 FY_9, nicknamed "Easterbunny") and Haumea (originally (136108) 2003 EL_{61}, nicknamed "Santa"), both announced on July 29, 2005, are larger still. Other objects, such as 28978 Ixion (discovered in 2001) and 20000 Varuna (discovered in 2000) measure roughly 500 km (311 mi) across.[2]

Pluto

The discovery of these large KBOs in similar orbits to Pluto led many to conclude that, bar its relative size, Pluto was not particularly different from other members of the Kuiper belt. Not only did these objects approach Pluto in size, but many also possessed satellites, and were of similar composition (methane and carbon monoxide have been found both on Pluto and on the largest KBOs).[2] Thus, just as Ceres was considered a planet before the discovery of its fellow asteroids, some began to suggest that Pluto might also be reclassified.

The issue was brought to a head by the discovery of Eris, an object in the scattered disc far beyond the Kuiper belt, that is now known to be 27 percent more massive than Pluto.[78] In response, the International Astronomical Union (IAU), was forced to define what a planet is for the first time, and in so doing included in their definition that a planet must have "cleared the neighbourhood around its orbit."[79] As Pluto shared its orbit with so many KBOs, it was deemed not to have cleared its orbit, and was thus reclassified from a planet to a member of the Kuiper belt.

Though Pluto is the largest KBO, a number of objects outside the Kuiper belt which may have begun their lives as KBOs are larger. Eris is the most obvious example, but Neptune's moon Triton, which, as explained above, is probably a captured KBO, is also larger than Pluto.

As of 2008, only five objects in the Solar System, Ceres, Pluto, Eris, Makemake and Haumea, are considered dwarf planets. However, a number of other Kuiper belt objects are also large enough to be spherical and could be classified as dwarf planets in the future.[80]

Satellites

Of the four largest TNOs, three (Eris, Pluto, and Haumea) possess satellites, and two have more than one. A higher percentage of the larger KBOs possess satellites than the smaller objects in the Kuiper belt, suggesting that a different formation mechanism was responsible.[81] There are also a high number of binaries (two objects close enough in mass to be orbiting "each other") in the Kuiper belt. The most notable example is the Pluto-Charon binary, but it is estimated that around 11 percent of KBOs exist in binaries.[82]

Exploration

Artist's conception of *New Horizons* at Pluto

On January 19, 2006 the first spacecraft mission to explore the Kuiper belt, *New Horizons,* was launched. The mission, headed by Alan Stern of the Southwest Research Institute, will arrive at Pluto on July 14, 2015, and, circumstances permitting, will continue on to study another as-yet undetermined KBO. Any KBO chosen will be between 25 and 55 miles (40 to 90 km) in diameter and, ideally, white or grey, to contrast with Pluto's reddish colour.[83] John Spencer, an astronomer on the *New Horizons* mission team, says that no target for a post-Pluto Kuiper belt encounter has yet been selected, as they are awaiting data from the Pan-STARRS survey project to ensure as wide a field of options as possible.[84] The Pan-STARRS project, partially operational since May 2010,[85] will, when fully online, survey the entire sky with four 1.4 gigapixel digital cameras to detect any moving objects, from near-earth objects to KBOs.[86] To speed up the detection process, the New Horizons team established Ice Hunters, a citizen science project that allows members of the public to participate in the search for suitable KBO targets.[87] [88] [89]

Other Kuiper belts

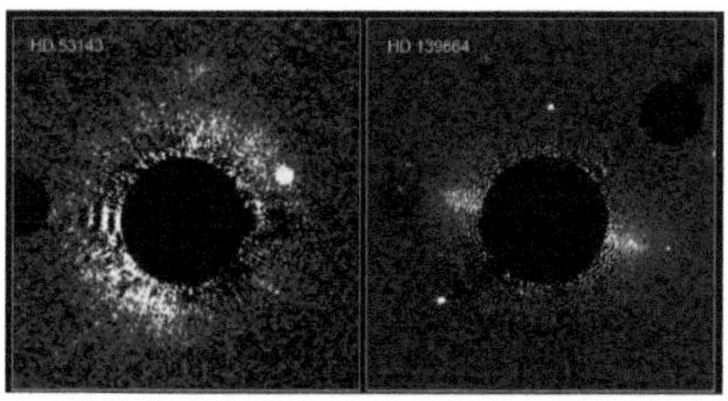

The debris disks around two stars (HD 139664 and HD 53143). The black central circle is produced by the camera's coronagraph, which hides the central star to allow the much fainter disks to be seen.

By 2006, astronomers had resolved dust disks believed to be Kuiper belt-like structures around nine stars other than the Sun. They appear to fall into two categories: wide belts, with radii of over 50 AU, and narrow belts (like our own Kuiper belt) with radii of between 20 and 30 AU and relatively sharp boundaries.[90] Beyond this, 15–20% of solar-type stars have an observed infrared excess which is believed to indicate massive Kuiper belt-like structures.[91] Most known debris discs around other stars are fairly young, but the two images on the right, taken by the Hubble Space Telescope in January 2006, are old enough (roughly 300 million years) to have settled into stable configurations. The left image is a "top view" of a wide belt, and the right image is an "edge view" of a narrow belt.[90] [92] Supercomputer simulations of dust in the Kuiper belt suggest that when it was younger, it may have resembled the narrow rings seen around younger stars.[93]

See also

- List of trans-Neptunian objects
- List of plutoid candidates

Notes

[1] Alan Stern; Colwell, Joshua E. (1997). "Collisional Erosion in the Primordial Edgeworth-Kuiper Belt and the Generation of the 30–50 AU Kuiper Gap". *The Astrophysical Journal* **490** (2): 879–882. Bibcode 1997ApJ...490..879S. doi:10.1086/304912.

[2] Audrey Delsanti and David Jewitt. "The Solar System Beyond The Planets" (http://web.archive.org/web/20070925203400/http://www.ifa.hawaii.edu/faculty/jewitt/papers/2006/DJ06.pdf). *Institute for Astronomy, University of Hawaii.* Archived from the original (http://www2.ess.ucla.edu/~jewitt/papers/2006/DJ06.pdf) on September 25, 2007. . Retrieved March 9, 2007.

[3] Krasinsky, G. A.; Pitjeva, E. V.; Vasilyev, M. V.; Yagudina, E. I. (July 2002). "Hidden Mass in the Asteroid Belt". *Icarus* **158** (1): 98–105. Bibcode 2002Icar..158...98K. doi:10.1006/icar.2002.6837.

[4] Johnson, Torrence V.; and Lunine, Jonathan I.; *Saturn's moon Phoebe as a captured body from the outer Solar System*, Nature, Vol. 435, pp. 69–71

[5] Craig B. Agnor & Douglas P. Hamilton (2006). "Neptune's capture of its moon Triton in a binary-planet gravitational encounter" (http://web.archive.org/web/20070621182809/http://www.es.ucsc.edu/~cagnor/papers_pdf/2006AgnorHamilton.pdf). *Nature*. Archived from the original (http://www.es.ucsc.edu/~cagnor/papers_pdf/2006AgnorHamilton.pdf) on June 21, 2007. . Retrieved June 20, 2006.

[6] Jewitt, David; Luu, Jane (1993). "Discovery of the candidate Kuiper belt object 1992 QB1". *Nature* **362** (6422): 730. Bibcode 1993Natur.362..730J. doi:10.1038/362730a0.

[7] David Jewitt. "Kuiper Belt Page" (http://www2.ess.ucla.edu/~jewitt/kb.html). . Retrieved October 15, 2007.

[8] Harold F. Levison, Luke Donnes (2007). "Comet Populations and Cometary Dynamics". In Lucy Ann Adams McFadden, Paul Robert Weissman, Torrence V. Johnson. *Encyclopedia of the Solar System* (2nd ed.). Amsterdam; Boston: Academic Press. pp. 575–588. ISBN 0-12-088589-1.

[9] The literature is inconsistent in the use of the phrases "scattered disc" and "Kuiper belt". For some, they are distinct populations; for others, the scattered disk is part of the Kuiper belt. Authors may even switch between these two uses in a single publication.Weissman and Johnson, 2007, *Encyclopedia of the solar system*, footnote p. 584 In this article, the scattered disk will be considered a separate population from the Kuiper belt.

[10] Gérard FAURE (2004). "Description of the System of Asteroids as of May 20, 2004" (http://web.archive.org/web/20070529003558/http://www.astrosurf.com/aude/map/us/AstFamilies2004-05-20.htm). Archived from the original (http://www.astrosurf.com/aude/map/us/AstFamilies2004-05-20.htm) on May 29, 2007. . Retrieved June 1, 2007.

[11] "What is improper about the term "Kuiper belt"? (or, Why name a thing after a man who didn't believe its existence?)" (http://www.icq.eps.harvard.edu/kb.html). *International Comet Quarterly*. . Retrieved October 24, 2010.

[12] John Davies (2001). *Beyond Pluto: Exploring the outer limits of the solar system.* Cambridge University Press. xii.

[13] Davies, p. 2

[14] David Jewitt. "WHY "KUIPER" BELT?" (http://www2.ess.ucla.edu/~jewitt/kb/gerard.html). *University of Hawaii*. . Retrieved June 14, 2007.

[15] Davies, p. 14

[16] Rao, M. M. (1964). "Decomposition of Vector Measures" (http://www.pnas.org/cgi/reprint/51/5/711.pdf). *Proceedings of the National Academy of Sciences* **51** (5): 771. Bibcode 1964PNAS...51..771R. doi:10.1073/pnas.51.5.771. .

[17] CT Kowal, W Liller, BG Marsden; Liller; Marsden (1977). "The discovery and orbit of /2060/ Chiron". *In: Dynamics of the solar system; Proceedings of the Symposium* **81**: 245. Bibcode 1979IAUS...81..245K.

[18] JV Scotti, DL Rabinowitz, CS Shoemaker, EM Shoemaker, DH Levy, TM King, EF Helin, J Alu, K Lawrence, RH McNaught, L Frederick, D Tholen, BEA Mueller; Rabinowitz; Shoemaker; Shoemaker; Levy; King; Helin; Alu et al. (1992). "1992 AD". *IAU Circ.* **5434**: 1. Bibcode 1992IAUC.5434....1S.

[19] Horner, J.; Evans, N.W.; Bailey, M. E. (2004). "Simulations of the Population of Centaurs I: The Bulk Statistics". *MNRAS* **354** (3): 798–810. arXiv:astro-ph/0407400. Bibcode 2004MNRAS.354..798H. doi:10.1111/j.1365-2966.2004.08240.x.

[20] Davies p. 38

[21] David Jewitt (2002). "From Kuiper Belt Object to Cometary Nucleus: The Missing Ultrared Matter". *The Astronomical Journal* **123** (2): 1039–1049. Bibcode 2002AJ....123.1039J. doi:10.1086/338692.

[22] Oort, J. H. (1950). "The structure of the cloud of comets surrounding the Solar System and a hypothesis concerning its origin". *Bull. Astron. Inst. Neth.* **11**: 91. Bibcode 1950BAN....11...91O.

[23] Davies p. 39

[24] JA Fernandez (1980). "On the existence of a comet belt beyond Neptune". *Monthly Notices of the Royal Astronomical Society* **192**: 481. Bibcode 1980MNRAS.192..481F.

[25] M. Duncan, T. Quinn, and S. Tremaine (1988). "The origin of short-period comets". *Astrophysical Journal* **328**: L69. Bibcode 1988ApJ...328L..69D. doi:10.1086/185162.

[26] Davies p. 191

[27] Davies p. 50

[28] Davies p. 51

[29] Davies pp. 52, 54, 56

[30] Davies pp. 57, 62

[31] Davies p. 65

[32] BS Marsden; Jewitt; Marsden (1993). "1993 FW". *IAU Circ.* **5730**: 1. Bibcode 1993IAUC.5730....1L.

[33] Davies p. 199

[34] Kathryn Hansen (June 7, 2005). "Orbital shuffle for early solar system" (http://www.geotimes.org/june05/WebExtra060705.html). *Geotimes*. . Retrieved August 26, 2007.

[35] Thommes, E. W.; Duncan, M. J.; Levison, H. F. (2002). "The Formation of Uranus and Neptune among Jupiter and Saturn". *The Astronomical Journal* **123** (5): 2862. arXiv:astro-ph/0111290. Bibcode 2002AJ....123.2862T. doi:10.1086/339975.

[36] Renu Malhotra (1994). "Nonlinear Resonances in the Solar System". *Physica D: Nonlinear Phenomena* **77**: 289. arXiv:chao-dyn/9406004. Bibcode 1994PhyD...77..289M. doi:10.1016/0167-2789(94)90141-4.

[37] Lovett, Rick (2010). "Kuiper Belt may be born of collisions". *Nature*. doi:10.1038/news.2010.522.

[38] M. C. De Sanctis, M. T. Capria, and A. Coradini (2001). "Thermal Evolution and Differentiation of Edgeworth-Kuiper Belt Objects". *The Astronomical Journal* **121** (5): 2792–2799. Bibcode 2001AJ....121.2792D. doi:10.1086/320385.

[39] "Discovering the Edge of the Solar System" (http://web.archive.org/web/20080305191925/http://www.americanscientist.org/template/AssetDetail/assetid/25723/page/2;jsessionid=aaa5LVF0). *American Scientists.org*. 2003. Archived from the original (http://www.americanscientist.org/template/AssetDetail/assetid/25723/page/2;jsessionid=aaa5LVF0) on March 5, 2008. . Retrieved June 23, 2007.

[40] Michael E. Brown, Margaret Pan (2004). "The Plane of the Kuiper Belt". *The Astronomical Journal* **127** (4): 2418–2423. Bibcode 2004AJ....127.2418B. doi:10.1086/382515.

[41] Jean-Marc Petit, Alessandro Morbidelli, Giovanni B. Valsecchi (1998). "Large Scattered Planetesimals and the Excitation of the Small Body Belts" (http://www.obs-nice.fr/morby/papers/6166a.pdf). . Retrieved June 23, 2007.

[42] Jonathan Lunine (2003). "The Kuiper Belt" (http://www.gsmt.noao.edu/gsmt_swg/SWG_Apr03/The_Kuiper_Belt.pdf). . Retrieved June 23, 2007.

[43] Dave Jewitt (2004). "CLASSICAL KUIPER BELT OBJECTS (CKBOs)" (http://web.archive.org/web/20070609094740/http://www.ifa.hawaii.edu/~jewitt/kb/kb-classical.html). Archived from the original (http://www2.ess.ucla.edu/~jewitt/kb/kb-classical.html) on June 9, 2007. . Retrieved June 23, 2007.

[44] P Murdin (2000). "Cubewano". *The Encyclopedia of Astronomy and Astrophysics*. Bibcode 2000eaa..bookE5403.. doi:10.1888/0333750888/5403. ISBN 0333750888.

[45] J. L. Elliot, S. D. Kern, K. B. Clancy, A. A. S. Gulbis, R. L. Millis, M. W. Buie, L. H. Wasserman, E. I. Chiang, A. B. Jordan, D. E. Trilling, and K. J. Meech (2004). "THE DEEP ECLIPTIC SURVEY: A SEARCH FOR KUIPER BELT OBJECTS AND CENTAURS. II. DYNAMICAL CLASSIFICATION, THE KUIPER BELT PLANE, AND THE CORE POPULATION" (http://alpaca.as.arizona.edu/~trilling/des2.pdf). . Retrieved June 23, 2007.

[46] "Naming of astronomical objects: Minor planets" (http://www.iau.org/public_press/themes/naming/#minorplanets). *International Astronomical Union*. . Retrieved November 17, 2008.

[47] Harold F. Levison, Alessandro Morbidelli (2003). "The formation of the Kuiper belt by the outward transport of bodies during Neptune's migration" (http://www.obs-nice.fr/morby/stuff/NATURE.pdf). . Retrieved June 25, 2007.

[48] Alessandro Morbidelli (2005). "Origin and Dynamical Evolution of Comets and their Reservoirs". arXiv:astro-ph/0512256 [astro-ph].

[49] "List Of Transneptunian Objects" (http://www.minorplanetcenter.org/iau/lists/TNOs.html). *Minor Planet Center*. . Retrieved June 23, 2007.

[50] "Ixion" (http://ixion.eightplanets.net/). *eightplanets.net*. . Retrieved June 23, 2007.

[51] John Stansberry; Will Grundy; Mike Brown; Dale Cruikshank; John Spencer; David Trilling; Jean-Luc Margot (2007). "Physical Properties of Kuiper Belt and Centaur Objects: Constraints from Spitzer Space Telescope". arXiv:astro-ph/0702538 [astro-ph].

[52] Chiang et al. (2003). "Resonance Occupation in the Kuiper Belt: Case Examples of the 5:2 and Trojan Resonances". *The Astronomical Journal* **126** (1): 430–443. arXiv:astro-ph/0301458. Bibcode 2003AJ....126..430C. doi:10.1086/375207.

[53] Wm. Robert Johnston (2007). "Trans-Neptunian Objects" (http://www.johnstonsarchive.net/astro/tnos.html). . Retrieved June 23, 2007.

[54] Davies p. 104

[55] Davies p. 107

[56] E. I. Chiang and M. E. Brown (1999). "Keck Pencil-Beam Survey For Faint Kuiper Belt Objects" (http://www.gps.caltech.edu/~mbrown/papers/ps/kbodeep.pdf). . Retrieved July 1, 2007.

[57] G.M. Bernstein, D.E. Trilling, R.L. Allen, M.E. Brown, M. Holman and R. Malhotra (2004). "The Size Distribution of Trans-Neptunian Bodies" (http://www.gps.caltech.edu/~mbrown/papers/ps/bernstein.pdf). *The Astrophysical Journal* **128**: 1364. arXiv:astro-ph/0308467.

Bibcode 2004AJ....128.1364B. doi:10.1086/422919. .

[58] Michael Brooks (2007). "13 Things that do not make sense" (http://space.newscientist.com/article.ns?id=mg18524911.600). *NewScientistSpace.com*. . Retrieved June 23, 2007.

[59] Govert Schilling (2008). "The mystery of Planet X" (http://space.newscientist.com/article/mg19726381.600-the-mystery-of-planet-x.html). *New Scientist*. . Retrieved February 8, 2008.

[60] Stephen C. Tegler (2007). "Kuiper Belt Objects: Physical Studies". In Lucy-Ann McFadden *et al.*. *Encyclopedia of the Solar System*. pp. 605–620.

[61] Altwegg, K.; Balsiger, H.; Geiss, J. (1999). *Space Science Reviews* **90**: 3. Bibcode 1999SSRv...90....3A. doi:10.1023/A:1005256607402.

[62] David C. Jewitt & Jane Luu (2004). "Crystalline water ice on the Kuiper belt object (50000) Quaoar" (http://web.archive.org/web/20070621182808/http://www.ifa.hawaii.edu/~jewitt/papers/50000/Quaoar.pdf). Archived from the original (http://www2.ess.ucla.edu/~jewitt/papers/50000/Quaoar.pdf) on June 21, 2007. . Retrieved June 21, 2007.

[63] Dave Jewitt (2004). "Surfaces of Kuiper Belt Objects" (http://web.archive.org/web/20070609094911/http://www.ifa.hawaii.edu/~jewitt/kb/kb-colors.html). *University of Hawaii*. Archived from the original (http://www2.ess.ucla.edu/~jewitt/kb/kb-colors.html) on June 9, 2007. . Retrieved June 21, 2007.

[64] Jewitt, David; Luu, Jane (1998). "Optical-Infrared Spectral Diversity in the Kuiper Belt". *The Astronomical Journal* **115** (4): 1667. Bibcode 1998AJ....115.1667J. doi:10.1086/300299.

[65] Davies p. 118

[66] Jewitt, David C.; Luu, Jane X. (2001). "Colors and Spectra of Kuiper Belt Objects". *The Astronomical Journal* **122** (4): 2099. arXiv:astro-ph/0107277. Bibcode 2001AJ....122.2099J. doi:10.1086/323304.

[67] Brown, R. H.; Cruikshank, DP; Pendleton, Y; Veeder, GJ (1997). "Surface Composition of Kuiper Belt Object 1993SC". *Science* **276** (5314): 937–9. Bibcode 1997Sci...276..937B. doi:10.1126/science.276.5314.937. PMID 9163038.

[68] Brown, Michael E.; Blake, Geoffrey A.; Kessler, Jacqueline E. (2000). "Near-Infrared Spectroscopy of the Bright Kuiper Belt Object 2000 EB173". *The Astrophysical Journal* **543** (2): L163. Bibcode 2000ApJ...543L.163B. doi:10.1086/317277.

[69] Licandro; Oliva; Di MArtino (2001). "NICS-TNG infrared spectroscopy of trans-neptunian objects 2000 EB173 and 2000 WR106". *Astronomy and Astrophysics* **373** (3): L29. arXiv:astro-ph/0105434. Bibcode 2001A&A...373L..29L. doi:10.1051/0004-6361:20010758.

[70] Gladman, Brett; et al (05/April/2001). "The structure of the Kuiper belt". *Astronomical Journal* **122** (2): 1051–1066. Bibcode 2001AJ....122.1051G. doi:10.1086/322080.

[71] Lorenzo Iorio (2007). "Dynamical determination of the mass of the Kuiper Belt from motions of the inner planets of the Solar system". *Monthly Notices of the Royal Astronomical Society* **375** (4): 1311–1314. Bibcode 2007MNRAS.tmp...24I. doi:10.1111/j.1365-2966.2006.11384.x.

[72] Bernstein, G. M.; Trilling, D. E.; Allen, R. L.; Brown, K. E.; Holman, M.; Malhotra, R. (2004). "The size distribution of transneptunian bodies". *The Astronomical Journal* **128** (3): 1364–1390. arXiv:astro-ph/0308467. Bibcode 2004AJ....128.1364B. doi:10.1086/422919.

[73] l, Eric (January 13, 2010). "Hubble Finds Smallest Kuiper Belt Object" (http://www.meteoritesusa.com/meteorite-photos/hubble-finds-smallest-kuiper-belt-object/). *MeteoritesUSA*. . Retrieved January 16, 2011.

[74] "List Of Centaurs and Scattered-Disk Objects" (http://www.minorplanetcenter.org/iau/lists/Centaurs.html). *IAU: Minor Planet Center*. . Retrieved October 27, 2010.

[75] David Jewitt (2005). "The 1000 km Scale KBOs" (http://www2.ess.ucla.edu/~jewitt/kb/big_kbo.html). *University of Hawaii*. . Retrieved July 16, 2006.

[76] Craig B. Agnor & Douglas P. Hamilton (2006). "Neptune's capture of its moon Triton in a binary-planet gravitational encounter" (http://web.archive.org/web/20070621182809/http://www.es.ucsc.edu/~cagnor/papers_pdf/2006AgnorHamilton.pdf). *Nature*. Archived from the original (http://www.es.ucsc.edu/~cagnor/papers_pdf/2006AgnorHamilton.pdf) on June 21, 2007. . Retrieved October 29, 2007.

[77] DALE P. CRUIKSHANK (2004). *TRITON, PLUTO, CENTAURS, AND TRANS-NEPTUNIAN BODIES* (http://books.google.com/?id=MbmiTd3x1UcC&pg=PA421&dq=.+TRITON,+PLUTO,+CENTAURS,+AND+TRANS-NEPTUNIAN+BODIES). Springer. ISBN 978-1-4020-3362-9. . Retrieved June 23, 2007.

[78] Mike Brown (2007). "Dysnomia, the moon of Eris" (http://www.gps.caltech.edu/~mbrown/planetlila/moon/index.html). *CalTech*. . Retrieved June 14, 2007.

[79] "Resolution B5 and B6" (http://www.iau.org/static/resolutions/Resolution_GA26-5-6.pdf). International Astronomical Union. 2006. .

[80] "IAU Draft Definition of Planet" (http://www.iau.org/iau0601.424.0.html). *IAU*. 2006. . Retrieved October 26, 2007.

[81]

This citation will be automatically completed in the next few minutes. You can jump the queue or expand by hand (http://en.wikipedia.org/wiki/Template:cite_doi/_10.1086.2f501524_)

[82] Agnor, C.B.; Hamilton, D.P. (2006). "Neptune's capture of its moon Triton in a binary-planet gravitational encounter" (http://www.astro.umd.edu/~hamilton/research/reprints/AgHam06.pdf). *Nature* **441** (7090): 192–4. Bibcode 2006Natur.441..192A. doi:10.1038/nature04792. PMID 16688170. .

[83] "New Horizons mission timeline" (http://pluto.jhuapl.edu/mission/mission_timeline.php). *NASA*. . Retrieved March 19, 2011.

[84] Cal Fussman (2006). "The Man Who Finds Planets" (http://discovermagazine.com/2006/may/cover/article_view?b_start:int=3&-C=). *Discover magazine*. . Retrieved August 13, 2007.

[85] Institute for Astronomy, University of Hawai (2010). "PS1 goes Operational and begins Science Mission, May 2010" (http://pan-starrs.ifa.hawaii.edu/public/project-status/PS1 operational.html). . Retrieved August 30, 2010.

[86] "Pan-Starrs: University of Hawaii" (http://pan-starrs.ifa.hawaii.edu/public/home.html). 2005. . Retrieved August 13, 2007.
[87] "Ice Hunters web site" (http://www.icehunters.org). Zooniverse.Org. . Retrieved July 8, 2011.
[88] "Citizen Scientists: Discover a New Horizons Flyby Target" (http://solarsystem.nasa.gov/news/display.cfm?News_ID=37726). NASA. Jun 21, 2011. . Retrieved August 23, 2011.
[89] Lakdawalla, Emily (June 21, 2011). "The most exciting citizen science project ever (to me, anyway)" (http://planetary.org/blog/article/00003073/). The Planetary Society. . Retrieved August 31, 2011.
[90] Kalas, Paul; Graham, James R.; Clampin, Mark C.; Fitzgerald, Michael P. (2006). "First Scattered Light Images of Debris Disks around HD 53143 and HD 139664". *The Astrophysical Journal* **637**: L57. arXiv:astro-ph/0601488. Bibcode 2006ApJ...637L..57K. doi:10.1086/500305.
[91] Trilling, D. E.; Bryden, G.; Beichman, C. A.; Rieke, G. H.; Su, K. Y. L.; Stansberry, J. A.; Blaylock, M.; Stapelfeldt, K. R.; Beeman, J. W.; Haller, E. E. (February 2008). "Debris Disks around Sun-like Stars". *The Astrophysical Journal* **674** (2): 1086–1105. Bibcode 2008ApJ...674.1086T. doi:10.1086/525514.
[92] "Dusty Planetary Disks Around Two Nearby Stars Resemble Our Kuiper Belt" (http://hubblesite.org/newscenter/archive/releases/2006/05/image/a). 2006. . Retrieved July 1, 2007.
[93] Kuchner, M. J.; Stark, C. C. (2010). "Collisional Grooming Models of the Kuiper Belt Dust Cloud". *The Astronomical Journal* **140** (4): 1007–1019. Bibcode 2010AJ....140.1007K. doi:10.1088/0004-6256/140/4/1007.

References

External links and data sources

- Dave Jewitt's page @ UCLA (http://www2.ess.ucla.edu/~jewitt/kb.html)
  - The belt's name (http://www2.ess.ucla.edu/~jewitt/kb/gerard.html)
- Ice Hunters (http://www.icehunters.org/), a citizen science project to help discover a KBO to serve as a followup *New Horizons* mission target after Pluto
- List of short period comets by family (http://www.physics.ucf.edu/~yfernandez/cometlist.html)
- Kuiper Belt Profile (http://solarsystem.nasa.gov/planets/profile.cfm?Object=KBOs) by NASA's Solar System Exploration (http://solarsystem.nasa.gov)
- The Kuiper Belt Electronic Newsletter (http://www.boulder.swri.edu/ekonews/)
- Wm. Robert Johnston's TNO page (http://www.johnstonsarchive.net/astro/tnos.html)
- Minor Planet Center: Plot of the Outer Solar System (http://www.minorplanetcenter.org/iau/lists/OuterPlot.html), illustrating Kuiper gap
- Website of the International Astronomical Union (http://www.iau.org/) (debating the status of TNOs)
- XXVIth General Assembly 2006 (http://www.astronomy2006.com)
- nature.com article: diagram displaying inner solar system, Kuiper Belt, and Oort Cloud (http://www.nature.com/nature/journal/v424/n6949/fig_tab/nature01725_F1.html), taken from Alan Stern, S. (2003). "The evolution of comets in the Oort cloud and Kuiper belt". *Nature* **424** (6949): 639–42. doi:10.1038/nature01725. PMID 12904784.
- SPACE.com: Discovery Hints at a Quadrillion Space Rocks Beyond Neptune (http://www.space.com/scienceastronomy/060814_tno_found.html) (Sara Goudarzi) August 15, 2006 06:13 am ET
- The Outer Solar System (http://www.astronomycast.com/astronomy/episode-64-pluto-and-the-icy-outer-solar-system/) Astronomy Cast episode No. 64, includes full transcript.
- The Kuiper belt (http://365daysofastronomy.org/2009/12/08/december-8th-what-is-the-kuiper-belt/) at 365daysofastronomy.org

Article Sources and Contributors

Asteroid *Source*: http://en.wikipedia.org/w/index.php?title=Asteroid *Contributors*: (, -- April, 10metreh, 129.128.164.xxx, 129.128.90.xxx, 15lsoucy, 1981willy, 1984, 1wonjae, 23skidoo, 28421u2232nfenfcenc, 2help, 3morand, 95j, A-giau, A4, AATTW, ABF, ACSE, AVand, Abc518, Academic Challenger, Acalamari, Access Denied, Accurizer, Adam Keller, Adambro, Addps4cat, AdjustShift, Adnergje, Ageekgal, Aggie747, Ahoerstemeier, Ajitem49, Ajraddatz, Alan R. Fisher, Alansohn, Ale jrb, Aleenf1, Alex.muller, Alexf, AlexiusHoratius, Alfio, Allstarecho, AllyUnion, Alsandro, Amazins490, Amire80, Andonic, Andre Engels, AndrewBuck, Andy4789, AngelOfSadness, Angela, AnimAlu, Ann Stouter, Anna512, Antandrus, Antoni Barau, Anville, Apparition11, Apple K, Arcman, Arjun01, Arkuat, Aster2, Asteroidz R not planetz, Astrobiologist, Ato, Attackpig, Avicennasis, Ayelie, Aznrules, Backslash Forwardslash, Badgernet, Banzai314, Bathanyucolkol, Beemer69, Beland, Ben-Zin, Ben.davis35, Benhocking, Bercziszani, Bevo, Bihco, BilCat, Bill37212, Billymac00, Blanchardb, Bluemask, Bob f it, Bobblewik, Bobizuruncle, Bobo192, Bogey97, Bokac1984, Bomac, Bongwarrior, Brandon, Brion VIBBER, Brockert, Bryan Derksen, Bsimmons666, CBDunkerson, CWenger, CWii, Callumw1, Calvin 1998, Can't sleep, clown will eat me, CapitalR, Captain Yesterday, Captain-tucker, Captainbeefart, Carinemily, Carmichael95, Carson69, Casmith 789, Catapult, Catttttt2, Celarnor, Ceranthor, Charles Matthews, Charliesface, CharlotteWebb, Chase me ladies, I'm the Cavalry, Chesnok, ChiZeroOne, Chris goulet, Chuck02, Chunkylefunga, Ck lostsword, Ckatz, Ckruschke, Clovis Sangrail, Cmapm, CommonsDelinker, Conversion script, CoolbeansofGR, Coolkid1001, Courcelles, Craigboy, Creidieki, Cryptic, Css, Curps, CzarB, DARTH SIDIOUS 2, DMacks, DVD R W, DW declare, DaL33T, Dam sens, Danakil, Daniel, DanielCD, DanwWiki, Daranak, Dartoe, David0811, DavidLevinson, DazBoz, Dcoetzee, Deagle AP, Dekaels, Dendodge, Deor, Deuar, Dffgd, Diderot, Dina, Discospinster, Divebomb, Dman159753, Dr. Blofeld, DrGatsby1962, Dragon of the Pants, Drbogdan, Dreish, Drift chambers, Drmies, Dstrike224, Duoduoduo, Dwadejames03, Dysepsion, E. Fokker, ERcheck, Editor at Large, Eeekster, Eggoeater, Ejosse1, El C, El Shaday, Eleassar, Emerson7, Emijrp, Emote, Emurphy42, Enemyofthestate, EoGuy, Epbr123, Eric-Wester, Ericann726, Erik9, Etacar11, Evil Monkey, Evil saltine, Evvyleeds, Excirial, Eyrian, FM, Fabiform, Fabiob, Faker12345, Falcon8765, Fartherred, Femto, Ferbtastic, Finlay McWalter, Finngall, Firsfron, Fisherama1, Fld300b, Flex Flint, Flyguy649, FlyingToaster, Folajimi, Foobaz, Fosnez, Fountains of Bryn Mawr, Fourohfour, Freakofnurture, Fredrik, Fritzey97, Fryergreg, Furrykef, Gaius Cornelius, Giftlite, Gilliam, Glen, Glenn, Globalsolidarity, Gogo Dodo, Golbez, Golgofrinchian, Gomada, GorillaWarfare, Graham87, Greenmango, GregorB, Greswik, Grstain, Gtg204y, Gtrmp, Guanaco, Gurch, Gwillhickers, H1nkles, Hackerrye, Halfway to never, Hamtechperson, Hanacy, Hanberke, Happysailor, Hapsiainen, Hdt83, Helpmet, HenryLi, Hiddekel, Hiddenfromview, Hike395, Holoeconomics, HorsePunchKid, Hydrogen Iodide, Icairns, Immunize, Insanity Incarnate, Ioscius, Iridescent, Island Monkey, Ispy1981, J 1982, J appleseed2, J.delanoy, JDG, JForget, JFreeman, JHunterJ, JYolkowski, Ja 62, Jacob1207, Jadtnr1, Jake838, Jaksmata, JamieS93, Jantzen is cool, Jared Preston, JavierMC, Jiddisch, Jmrowland, Jn1611, Joelwest, John Cumbers, John Deming, John254, Jojhutton, Jon Harald Søby, Jonathunder, JorisvS, Jose77, Josh-Levin@ieee.org, Josh70597, Jpk, Jpo, Jusdafax, JustAddPeter, Jwrosenzweig, Jyril, Jóna Þórunn, Kazvorpal, Kborer, Keilana, Kelly Martin, Kenohki, Kevyn, Kheider, Kier07, Kiiwiii, King of Hearts, Kingpin13, Kirill Lokshin, Kjoonlee, KleinKlio, Kozawa, Kubigula, Kudret abi, Kurtan, Kuru, Kwamikagami, L337 kybldmstr, LALALA26, La Parka Your Car, Lamb99, Latics, Latitudinarian, Leevanjackson, Lesnail, Lexor, Lifefeed, Lightmouse, Lightyear69, Liken to a glove, LilliPuppy, Limonns, Ling.Nut, Logan, Lolihatepizza, Looxix, Loren.wilton, Loresayer, Lowe4091, Luckybeargod, Luna Santin, Lupin, MATRIX, MER-C, MONGO, MacroDaemon, Maelnuneb, Magister Mathematicae, Majorclanger, Mandarax, Mani1, Manmobile, MarcK, Marek69, Mark Foskey, Martarius, Martin451, Mas 18 dl, Masonbarge, Master Jay, Materialscientist, Mathewignash, Matt Yeager, Matubo, Maury Markowitz, Mav, MavrickWeirdo, Mbc362, McSly, Mechanical digger, Mejor Los Indios, Menchi, Mephistophelian, Metalb, Metebelis, Mgmirkin, Michael Hardy, Micky kwoskie, Midnightcomm, Mike s, MikeCapone, MikeLynch, Mikenorton, Minesweeper, Minimac, MithrandirAgain, Mjuancarlo, Moomoomoo, Morwen, Moverton, Mrholybrain, Mtnerd, Muijz, Murgh, Murlough23, Mxn, Myanw, Mygerardromance, NAHID, Nakon, Narcosis17, NawlinWiki, Neier, Nephelin, Nergaal, Netoholic, Newsaholic, Nick UA, Nihiltres, NiiSSA, Njaelkies Lea, Noclevername, Novangelis, Nubiatech, Ocaasi, Oleg Alexandrov, Olegwiki, Omicronpersei8, Orangemarlin, OwenX, Oxymoron83, PTJoshua, PTSE, Patanet7, Patmic2502, Patrick, Paul H., Pauli133, Pedro, Perey, Pgan002, Phantomsteve, Philip Trueman, Pi@k, Pickom, Pinethicket, Piolinfax, Pit, Pizza Puzzle, PoeticVerse, Polkadotbaby0772, Polyamorph, Ponder, Porterjoh, Powered, Prashanthns, Pseudomonas, Python eggs, Q21op, Q43, Quadell, Quantpole, Quuxplusone, RFerreira, RJHall, RP88, RUL3R, RadicalOne, Rajabhattacharya, Ramesh Ramaiah, RandomCritic, Ravidreams, RazorICE, Razorflame, Rchandra, Reaper Eternal, RedWolf, Res2216firestar, Rettetast, RexNL, Reyk, Rich Farmbrough, Rivajunkie, Rjwilmsi, Rmhermen, RockMFR, Roentgenium111, Ron Ritzman, Rorschach, Rothorpe, Rubicon, Ruslik0, Rwflammang, Ryjaz, SJP, ST47, SU Linguist, Salmar, Sandhuharpreet, Sang'gre Habagat, Sapphic, Sardanaphalus, Saros136, Satori, SaveThePoint, Saxsux, SchfiftyThree, Schneelocke, SchnitzelMannGreek, Sciurinæ, Sdornan, Senator Palpatine, Serendipodous, Serpentinite, Seth Ilys, Shadowjams, Shalom Yechiel, Shanes, Shimgray, Shirifan, Shotgun5559, Shushruth, SiliconCerebrate, SkerHawx, Skier Dude, Skizzik, SkyWalker, Slicedoranges, Slightsmile, Slon02, Snowolf, Soccer dude 07, Sodium, Sole Soul, Some jerk on the Internet, Something14, Soonhuat95, Southern Illinois SKYWARN, Sp82, Spacepotato, Spartaz, Spartenxiv, Spencer, Spitfire, Standunne, Stealingitstrongly, Stemonitis, Stephenchou0722, Steven Zhang, StitchRFC, Stormwriter, Sturmde, Stwalkerster, Subversive.sound, SummerPhD, SuperHamster, Superbeecat, Susan714, Sushi, Svetovid, Swatjester, Synchronism, Tailpig, Taxman, Telescopi, Template namespace initialisation script, That Guy, From That Show!, The Enlightened, The High Fin Sperm Whale, The Phoenix, The Singing Badger, The Thing That Should Not Be, TheGWO, TheKoG, Theda, Thestupidguy, ThreeBlindMice, Thue, Tide rolls, TigerShark, Tillwe, Timir2, Timrollpickering, Timwi, Titoxd, Tjkiller555, Tom Peters, TomasBat, Tombomp, Topgearstopviewer, TransUtopian, Trash boy, Tree Biting Conspiracy, Trekphiler, Tresiden, Tschild, Tyleradkins31, Ultra two, Uni223344, Untifler, Unyoyega, Uppland, Urhixidur, Vansh9988, Vary, Vaġinator, Vgranucci, Vianello, Voxpuppet, Vrenator, Vsmith, Wackywace, Waltpohl, Waninge, Ward20, Wayne Hardman, Wayward, Werididiot, Werockthescienceclass, Westwood25, Wgungfu, Whitepaw, Why Not A Duck, Wiki alf, Wikiborg, Wikipelli, Wildyoda, Willhsmit, Willshark, WilyD, Wimt, Wipsenade, Wjfox2005, Wknight94, Wolfkeeper, WolfmanSF, Wwheaton, X relentless x, XJamRastafire, Xanucia, Xav71176, Xeno, Xerxes314, Xn44, Yamamoto Ichiro, Yoguys021298, Youknowandy, Z4ngetsu, Zhou Yu, Zidane tribal, Zsinj, Zundark, Zzzzz, ^demon, Île flottante, Саша Стефановић, Шугуан, 1374 anonymous edits

Asteroid_belt *Source*: http://en.wikipedia.org/w/index.php?title=Asteroid_belt *Contributors*: 1981willy, 21655, 2T, A-giau, Abbalkea2010, Ace45954, Adam Keller, Aesopos, Ahoerstemeier, Aiken drum, Airplaneman, Alansohn, AlexiusHoratius, Alfio, Alsandro, Amara, Anarchist42, Andrewlp1991, Antandrus, Apeman, Aponar Kestrel, Arsia Mons, Art LaPella, Arz1969, Avenue, AxelBoldt, Baa, Bender235, Bewildebeast, Bfissa, Bigger digger, BlastOButter42, Bobymeltics, Bongwarrior, Bryan Derksen, Burzmali, CanadianLinuxUser, Casliber, CharlesC, Cheat2win, Chesnok, ChristTrekker, ChristopherWillis, CielProfond, CirrusHead, Cjc13, Ckatz, Closedmouth, Coalisnotcools, Colonies Chris, Crash Underride, Crazymonkey1123, DABADABADAI, Dabomb87, Davewild, Dawnseeker2000, Delrayva, Denisarona, Deon, Deor, Deuar, Discospinster, Donarreiskoffer, Dougano, Dsparks000, Dspitzle, Dune010, Dwcarless, EarnestyEternity, Eccles999174, Ed Poor, Editore99, Edward130603, Egil, El C, El Cid, Emurphy42, Enviroboy, Epbr123, Everyking, Excirial, Extra999, FF2010, Falcon8765, Fcn, Felyza, Femto, Fjörgynn, Flyguy649, Folajimi, Fotaun, FrankCarroll, Frenchgizmo, Furrykef, Gaius Cornelius, Gap9551, Garion96, Geeoharee, Gene Nygaard, Geoff Plourde, George100, Gilliam, Gogo Dodo, Goudzovski, Grahamec, Gtg204y, Gurch, GuyQuest, Gwillhickers, Hadseys, Hairy Dude, HalfShadow, Hamiltonstone, Harry, Hashar, HatriX, Hda3ku, Hdt83, Headbomb, HenryLi, Herostratus, Hibernian, Hike395, Hmains, Hope(N Forever), Hunnjazal, I like trains beetches, IVAN3MAN, IanOsgood, Ianare, Icairns, Ilikebakedziti, Incnis Mrsi, Iridescent, J.delanoy, JForget, JHunterJ, Jacj, JamesHoadley, Jaredroberts, Jcronen1, Jeff G., Jeremy Visser, Jivecat, Jnestorius, JocK, Joepearson, John Reaves, JorisvS, Jusdafax, Jusjih, Jwh, Ke6jjj, Kejo13, Kheider, Kintetsubuffalo, Kobrabones, Kristaga, Kuru, Kwamikagami, Kyng, Lakers, Lampman, Lanthanum-138, Latulla, Lavateraguy, Leor klier, Lhopitalified, Liamdaly620, Lightmouse, Lilly Ishappy, LonelyMarble, Lotu, Luk, Luokehao, MJ94, Manisero399, Marcus Qwertyus, Marek69, Marksurge, Marskell, Martinwilke1980, Matthewedwards, Maxamegalon2000, McGeddon, Meelar, Megalodon99, Melara..., Merovingian, Metafax1, Mgiganteus1, Michael Devore, Michael Hardy, Michaelas10, Mike s, Milonica, Moneo, Moocha, Morgan Leigh, MrRadioGuy, Mulad, Murgh, Muriel Gottrop, Murlough23, Mxn, Myanw, Myrrhlin, Narayanese, Nephelin, NerdyScienceDude, Nick C, Nick123, Nivix, NorwegianBlue, Numbers123---, Oldmanjenkins93, Oni Ookami Alfador, Ontyx, Orii, Ost316, P30Carl, Palica, Papa November, Pembers, Peyre, Pharos, Philip Trueman, Piano non troppo, Pie4all88, Pinethicket, Pizza Puzzle, Pmanderson, PocketDocket, Poppy, Potatoswatter, PowerCS, Praefectorian, Quackers074, Quendus, Qurq, RJHall, RP88, RSStockdale, RadicalOne, Ran, Randomblue, Razorflame, Reaper Eternal, Res2216firestar, Retodon8, Revolución, RexNL, Reywas92, Rhowryn, Rich Farmbrough, Richhoncho, Rjwilmsi, Rmhermen, RobertG, Rogermw, Ronhjones, Rorschach, RossPatterson, Rothorpe, Rst20xx, Ruby.red.roses, Ruslik0, Rwflammang, SERK12, SMSshark, Sardanaphalus, Saros136, Sbig10, SchuminWeb, Scog, ScottyBerg, ScottyBoy900Q, Sdornan, Seaphoto, Sengkang, Serendipodous, Shadowcheets, Shadowjams, Shirik, Shshshsh, Siafu, Sjock, Skizzik, Skysmith, Smartech, SnappingTurtle, Snowolf, Some jerk on the Internet, Something14, Springnuts, Squirepants101, Sry85, Stephen e nelson, Subash.chandran007, Sugarfish, SuperHamster, SuperSpy00bob, TRUTH316, Taed, Tbhotch, Tcncv, Tearlach, Tedickey, Template namespace initialisation script, Teog, TerraFrost, The High Fin Sperm Whale, The Rambling Man, The Tom, Thehelpfulone, Tide rolls, Tiptoety, Titus III, Tombombadil, Tpbradbury, TravisTX, Treisijs, Tuckerresearch, Ufviper, Uncle Dick, Urhixidur, Vanished User 0001, Viriditas, Visor, Vittsadaf, Vsmith, WRK, Waxcaptain45, Wayne Hardman, Wayne Slam, West-J, Wetman, Whatcanbrowndo, WhiteDragon, Wiki alf, Wiki libs, Wikisux, Wiklol, William Avery, Wimt, Wipsenade, Wolfkeeper, WolfmanSF, Worldtraveller, Wtmitchell, X201, Xoxojoccy, Xxgurlxx, Yosef1987, Zachary, Zamphuor, Zanejune1, Zidane tribal, Zé da Silva, ₀x, 626 anonymous edits

Asteroid_family *Source*: http://en.wikipedia.org/w/index.php?title=Asteroid_family *Contributors*: Alfvaen, AxelBoldt, BigDwiki, Catapult, CharlotteWebb, Daigaku2051, Deuar, Epolk, Fjörgynn, Hans van Deukeren, Jkl, Jpo, Kheider, Kwamikagami, Mboverload, Mindmatrix, Mwhiz, Njardarlogar, QuantumEngineer, RP88, Rich Farmbrough, Richhoncho, Rursus, Siafu, The Singing Badger, Urhixidur, VirtualDave, WinstonSmith, 20 anonymous edits

Solar_System *Source*: http://en.wikipedia.org/w/index.php?title=Solar_System *Contributors*: (, (aeropagitica), -- April, -Majestic-, 0, 129.128.164.xxx, 1297, 141.157.52.xxx, 162.83.144.xxx, 216.237.32.xxx, 23skidoo, 24.93.53.xxx, 2nd Piston Honda, 711groove, A-giau, A. Parrot, AEMoreira042281, Aaron McDaid, Abbalkea2010, Abrech, Abuk78, Acalamari, Acather96, Acebrock, Acroterion, Adam850, Adashiel, Addshore, Adi, AdiJapan, Admiral Roo, Adraeus, Aelffin, Aeon1006, Ageekgal, AgentPeppermint, Ahadland1234, Ahoerstemeier, Ahsasin8, Aitias, Ajdz, Ajo Mama, Alanbly, Ale jrb, Alegoo92, AlexanderHaas, Alexdw, Algebraist, Alienkishore, Allstarecho, Alpertron, Altenmann, Amos Han, Amreatsf4620, Amyrlin, Anarchist42, Anaxial, Andonic, Andre Engels, AndrewHowse, Andrewlaneyboi93, Andy120290, AndyZ, Andyjsmith, Animum, Anitanorm, Anomalocaris, AnonEMouse, Antandrus, Anthony Arrigo, AntiVan, Antiestablishment, Antonio Lopez, Anusface2000, Anyeverybody, Apers0n, Aquirata, Aragupta, Arbitrary username, Arce, Arctic.gnome, Ardric47, ArnoldReinhold, Art LaPella, Ashill, Ashleyy osaurus, Ashmoo, Asqwce, Aster2, Asteroidz R not planetz, Astrolog, Athenean, Ato, Aubadaurada, Ausir, AutoFire, AutoGeek, AvicAWB, AxelBoldt, AySz88, AzaToth, Azcolvin429, BRG, BRUTE, Babbage, Badmonkey0001, Bajibble, Bambuush, BankingBum, Baseballdude12321, Bassem18, Batintherain, Bayerischermann, Bazzargh, Bbatsell, Bcorr, BeardedBob, Beast of traal, Beland, Belovedfreak, Ben Webber, Ben261994, BenAveling, Bender235, Benhocking, Bennymgradz, Berek, Bergsten, Bethy12, Beyazid, Bhadani, Bharathivdevi, Big Bird, Bigger digger, Bill37212, Billie Pipier, BirdKr, Bkdafkup, Bkell, Blakut, Blaxthos, Blaz5678, Blehfu, Blindman shady, Blobglob, BlueEarth, BlytheG, Bob Hu, Bob rulz, Bobblewik, Bobo192, Boing! said Zebedee, Bomac, Bonadea, Bond81895, Bookworm2002, Bootstoots, BorgQueen, Borislav, Bowlhover, BozMo, Br77rino, Brian0918, Brianga, Brighterorange, Brion VIBBER, Brisvegas, Bruce1ee, Bruin69, Bryan Derksen, Bryan Seecrets, BryanG, BudMann9, Bugliarisi, Burrito, Burwellian, C0nanPayne, CWii, Cadiomals, Caid Raspa, Caltas, Can't sleep, clown will eat me, Canadian-Bacon, CanadianCaesar, Canderson7, Caravaca, CarbonUnit, Carlj7, Carnildo, Carultch, Catgut, Cderoose, Cenarium, Ceoil, Chairman S., Chantelle2004, Chaosdruid, CharlesC,

Chem-awb, Chermundy, Chesnok, Chessphoon, Chipskip, Cholmes75, ChrisGriswold, Christopherlin, Chriswiki, Chupon, Ciaccona, Cimex, Cityside Seraph, Civil Engineer III, Ckatz, Clarityfiend, ClockworkTroll, Clq, Cmann100, Cmapm, Cmglee, Coastside, Cocoaguy, Cocu, Coelacan, Cogito ergo sumo, Cohenna, Coleflinn, ColinJF, Colonies Chris, Cometstyles, CommonsDelinker, Comp25, Complex (de), Computer user786, Computerjoe, Confession0791, Connelly, Constructive editor, Conversion script, Corpx, Costalotta, Cosumel, CovenantD, Coviepresb1647, Cquan, Crash Underride, Crazycomputers, Cremepuff222, Crevaner, Crevox, Cryptk, Cue the Strings, Curps, Cuttycuttiercuttiest, Cwill19, Cwolfsheep, Cyber boy, Cyclopia, Cyde, Cyktsui, Cymruman, CynicalMe, Cyril715, DARTH SIDIOUS 2, DDima, DMacks, DRTllbrg, DV8 2XL, Dabomb87, Daman numba23, Damërung, Daniel Olsen, Danielboerman, DannyZ, Danski14, Dar-Ape, Darkkid24, DarrynJ, DarthCat, DarthSwim, Dat789, Davewild, DavidLevinson, Davodd, Dazza anderson1, De Administrando Imperio, DeadEyeArrow, Dean Harper, Deaþe gecweald, Debresser, Deckiller, Decumanus, Deeptrivia, Dekaels, Dekimasu, Delaszk, Delfindakila, Delldot, Demmy100, Dendoi, DenisMoskowitz, Dennis Valeev, Denny, DerHexer, Derek Balsam, Derek Ross, Desmond71, Deuar, DeusMP, Deviathan, Dexeillic, Dia^, Diderot, Dina, Dirac1933, Discospinster, Dismas, Diverman, Dlimeb, Dlohcierekim, DocWatson42, DocendoDiscimus, Dodo von den Bergen, Dogyo, Domthedude001, Don'tknowenough, Doradus, Dr. Submillimeter, DrKiernan, Dragons flight, Dragostanasie, Drat, Dreg743, Dreish, Drmies, Drumguy8800, Dummyacc, Duoduoduo, Dvandersluis, Dwaipayanc, Dwarfplanets.org.uk, Dwheeler, Dynamite12321, Dzordzm, E Pluribus Anthony, E Wing, EarthPerson, Ec5618, Ed de Jonge, Ed g2s, Ederiel, Edgar181, Editer in disguise cool, Edivorce, Eduardo Sellan III, Egyptianboy15223, Eiaschool, Eilthireach, El C, Elagatis, Elert, Elf, Elliskev, Emarsee, Emilio floris, Emokid0987654321234567890, Ems57fcva, Emufarmers, Enceladusgeysers, Enirac Sum, Enviroboy, Eob, Epbr123, Equal, Er Komandante, Erebus555, Ergzay, Eric Forste, Eric Wester, Erikhansson1, Erimus, Eroica, Esanchez7587, Etxrge, Evanspahr5, Evercat, Everyking, Evgeni Sergeev, Evil saltine, Exert, Explicit, Extra999, Extransit, FF2010, Facts707, Fanghong, Farosdaughter, Farry, FastLizard4, Fatladi, FayssalF, Fedayee, Felix Portier, Ferdiaob, Feydey, Feyrauth, Figures&Puck, Fjörgynn, Flata, Fluorhydric, Flyguy649, Fnfd, Foobar, Fotaun, Franamax, Frank Shearar, Freakofnurture, Fredrik, Freewayguy, Freiza667, FreplySpang, Frip1000, Fullerene, Futuristcorporation, GDonato, Gadfium, Gaijin Ninja, Gaius Cornelius, Galoubet, Gameseeker, Gary King, Garycompugeek, Gateman1997, Gdo01, Geckodude92, Gene Mocoletlet, Gene Nygaard, GenestealerUK, Geni, Geob42, Geoff.green, Geologyguy, Georgia guy, Gerbrant, Ghewgill, Gian-2, Giftlite, Gifðas, Giggy, Gilliam, Gilsinan, Gimboid13, Glass Sword, Glen, Glenn, Glenn L, Gloern, Gman124, Gnixon, Gnomon Kelemen, Gogo Dodo, GoingBatty, Golbez, GoldenTorc, Goodnightmush, Gorillaman0321, Gota 93, GraemeL, Grandrek, Greenmage801, Greentsunami, Gregconquest, Grika, Ground Zero, Grstain, Grunt, Gscshoyru, Gsklee, Gt, Gurch, Gurchzilla, Guyzero, Gwillhickers, H, H2g2bob, Hadal, Hakkahakkabazoom, HalfShadow, Halsteadk, HammerHeadHuman, Harperska, HarryAlffa, Harryboyles, Hayter, Haza-w, Hdt83, Headbomb, Heals33, Hello i am a guy, Henhellen, Heron, Hersfold, Hertz1888, Hevron1998, Hi332211, HiDrNick, HiLo48, Hibernian, Hike395, Hmartin, Hojimachong, HorsePunchKid, Hurricane Devon, Husond, Hut 8.5, Hyad, Hypercephalic, I know about this, I'm supah awesome, ILHI, Iamunknown, Ianblair23, Iantresman, Ibagli, Ibroadbent, Icairns, IcePuckScore, Icestorm815, IdLoveOne, Idaltu, Ideogram, Iliev, Ima van Daal, Insanity Incarnate, InternetMeme, Iridescent, Irishguy, Iron knife, IronGargoyle, Isoxyl, Ixfd64, J P, JBG, JBlover23, JD554, JFreeman, JHunterJ, JQF, JStarStar, JWSchmidt, Jagged 85, James D. McBride, Jan.Kamenicek, Janviermichelle, Jao, Japak, Jaranda, Jason One, Jason Palpatine, JeLuF, Jeandré du Toit, Jeffmedkeff, Jeremy Visser, Jeronimo, Jerry, Jevans14, Jfiling, Jfpierce, Jfullerw, Jhenderson777, Jim Fitzgerald, Jim bob jr, Jimp, Jinma, Jkelly, JoanneB, JodyB, Joeborder, Joelwest, Joffan, John Quincy Adding Machine, JohnCub, JohnDBuell, Johnismyboo, Johnuniq, Jojit fb, JonWee, Jonathan Drain, Jorfer, JorisvS, Joseph Solis in Australia, Josh Parris, Joshua368, Jpallan, Jpk, Jredmond, Jrethorst, Jrockley, Julesd, Jumbuck, Jusjih, JustAGal, Jyril, K, K-UNIT, KAMiKAZOW, Kaldari, Kanags, Karaku, Karozoa, KathrynLybarger, Katydidit, Kaustubh, Kayau, Kchishol1970, Kcordina, Keilana, Kelvinc, Keraunos, Kevin29303, Keyboard mouse, Khalid Mahmood, Kheider, Kiensvay, Kim Bruning, King Lopez, Kingturtle, Kirbytime, Kitch, Kjkolb, Kjoonlee, Kmsiever, Knowledge Seeker, KnowledgeOfSelf, KojieroSaske, Kortaggio, Kozuch, KrakatoaKatie, Krash, Krich, Krótki, Kubigula, Kukini, Kumpees gay, Kungfuadam, Kunguni, Kuru, Kush2, KuwarOnline, Kwamikagami, Kwekubo, Kyokpae, KyraVixen, L Kensington, LAX, LOL, La goutte de pluie, LachlanA, Lacrimosus, Lanthanum-138, Laserfire, Latitude0116, Laudaka, Leafyplant, Lectonar, Lee, Leevanjackson, Leevclarke, Legolost, Leigui, Leki, Leon7, Lethe, Lexicon, LifeStar, Liftarn, Lightmouse, Lights, Limbo socrates, Linnell, Liop666, Liyster, Lo2u, Lomn, LonelyMarble, Lookang, Looxix, Lord Patrick, Lordmontu, Love fass, Lradrama, LuckyBob76, Luckyluke, LuigiManiac, Lukaigeneration, LukeSurl, Lulujannings, Lumos3, Luna Santin, Lunokhod, Lupin, MER-C, MONGO, MacTire, Madgeek09, Madhav krishnan, Madonna Can, Magister Mathematicae, Majorly, Makron1n, Malo, ManCreamy, Manipe, Manisero399, Manjithkaini, MaraNeo127, Marcus Qwertyus, Mark Foskey, Mark Musante, MarkSutton, Markhurd, MarsRover, Marshallsumter, Martin benettis@hotmail.com, Martin.Budden, Mastalroh, Master Jay, MasterSci, Masterhomer, Materialscientist, Matt Crypto, Matt. P, Matthead, Matty456789, Maurreen, Mav, Maxim, Mayumashu, Mboverload, Mbz1, McLarenJAB, Mdd, Mdf, Megabyte73, Meidosemme, Melaen, Memaster3, Meneth, Meno25, MeowMedia, Mephiles602, Merovingian, Mets501, MetsFan76, Mfrontz, Michael Hinrichs, Michael Patrick, Michael Slone, Michael riber jorgensen, Michaelas10, Michaelbarreto, Michaelbusch, Michalis Famelis, Microtony, Midgrid, Midnightcomm, Mike Christie, Mike Doughney, Mike s, Mikebrand, Mild Bill Hiccup, Mimihitam, Minesweeper, Minna Sora no Shita, MisfitToys, Misheru, MisterSheik, Mjb, Mlm42, Moanzhu, Mobiusorg, Mode11bob9, Monedula, Monotonehell, Montana's Defender, Monz, MoonshineOHNO, Mooseman927, Morelanj, Mormegil, Morshem, Moverton, Mpatel, Mr Tan, Mr. Garrison, Mr. Lefty, Mred64, Mrexxx, Mschel, Msikma, Mstroeck, MuDavid, Mufka, Multifinder17, Munchynik, Murgh, Murlough23, Mwtoews, Mxn, Myanw, Myguy123, Myles Trundle, Myrrhlin, Mzajac, N-Man, Nabla, Nacnud298, Nadav12, Nakon, Nations United, NatureA16, NawlinWiki, Nbound, Nedim Ardoğa, Nelson50, Nemrac2, Neonumbers, Nephelin, NerdyScienceDude, Nergaal, Nesnad, Neutrality, Neverquick, Newone, Nfitz, NickBush24, NickW557, Nickshanks, Nigholith, Night Gyr, Nihiltres, Nikai, Nimur, Nivix, Nixer, No-Bullet, NoIdeaNick, NobodysPerfect312, Noctibus, Noetica, Noosfractal, NorCalHistory, Norn Guy, Northumbrian, Notinasnaid, Nova77, Nurg, Nuttycoconut, Oboebasson, Ohms law, Olathe, Oliphaunt, Oliver Pereira, Oliver202, Omicronpersei8, Once in a Blue Moon, Onco p53, One, Onorem, Opelio, Orangutan, OrbitOne, Oren0, Osmaston, Ottershrew, Ouishoebean, Oxymoron83, PFHLai, PJS102, PJtP, PMP126, Paine Ellsworth, Paliosun, Panoptical, Papajfd, Parachnidox, Patrick, Patrickwooldridge, Patstuart, Paul August, Paul-L, Paulcmnt, Paulkondratuklovesny, Pb30, Pcj, Pedro, Penubag, Pergamino, Perrysweitzer, Peruvianllama, Peter, Peter Shearan, Peter Winnberg, Peterfwhyte, Pgecaj, Pgk, Pgr94, Ph89, Pharaoh of the Wizards, Phe, Phil 1970, Philip Trueman, Phillip J, Philolexica, Phoenix2, Phoenix79, PhySusie, Pi@k, Pictureuploader, Pilatus, Pilotguy, Pinethicket, Pious7, Piper2000ca, Pizza Puzzle, Pjrm, Planetary, Plau, PoccilScript, Poiu555, Polly, Polonium, Poor Yorick, Poplard, Postmortemjapan, Potatoedrink, PresN, Primetime, Prodego, Prometheus7Unbound, Proxima Centauri, Psyche825, Puntori, PuzzletChung, Pyrokitsune777, Python eggs, Q2op, Q3op, Qmwne235, QuantumEleven, Quintote, Qurq, Qwasty, Qwyrxian, Qxz, R, RJHall, RP88, Rachel Pearce, RadicalPi, Raditzu, Raichu Trainer, Raistlin11325, Raj Fra, Rake, Rama's Arrow, RandomCritic, Randomblue, Rashaani, Rasmus Faber, Ratsbew, Raven4x4x, Raymondwinn, RazorICE, Rbj2001, Rc3784, Real editor, Rebecca, Redgolpe, Redsully, Reedbraden, Renesis, Rettetast, RexNL, Rhuaridh, Riana, Richard D. LeCour, RichardF, Rick Block, Rickyrab, Ringmaster56, RingtailedFox, Rizzoj, Rjwilmsi, Rmky87, Rnt20, RobHolbert, Robbie098, Robby, RobertG, Robomaeyhem, RockMFR, Roentgenium111, Roerlepel, RoflSloth, Ronhjones, Roo60, Roofus, Rorschach, RoryReloaded, Roscoe x, Rothorpe, RoyBoy, Rreagan007, Rror, Rubicon, Run!, Rune.welsh, Rursus, Ruslik0, Ruud Koot, Rwboa22, Ryan Paddy, Ryulong, SJP, SMC89, SOCCERchic2134, SP-KP, ST47, Sabrebd, Sacre, Salamander03, Salamander44, Salamurai, Saluyot, Salvio giuliano, Sam Korn, Sam12321, Samjohnson, Samojari, Samoojas, Sampo Torgo, Samy Merchi, Sanchom, Sanderling, Sango123, Saramae101, Sardanaphalus, Saros136, Savant13, Sbandrews, Scarvie, Schneelocke, SchuminWeb, Schzmo, Scienz Guy, Scipius, Sciurinæ, Scjessey, Scott Ritchie, Scottosborne, Scwlong, Sd31415, Sean D Martin, Sean higgy2007, Selket, Sengkang, Sennaista, Serendipitous, Serendipodous, Sfahey, Sgt Pinback, Shadow007, Shadowjams, Shanes, Shanew2, Shawn81, SheffieldSteel, Shehee, Sherool, Sherz, Shimgray, Shinjiman, Shoeofdeath, Sidasta, Sietse Snel, Sikon, Silivrenion, Simple implication, Siniset, Sinn, Sir Nicholas de Mimsy-Porpington, Siroxo, Sj, Sjakkalle, SkerHawx, Skizzik, SkyMachine, Slakr, Sleeping123, Slysplace, Smartech, Smartse, Smarty Pants786, Smige30, Snigbrook, Snowolf, Solipsist, Someguy1221, Something14, Soumya11, Soumyasch, Spaceman8815, Spacepotato, Spangineer, Sparky the Seventh Chaos, SparrowsWing, Spazturtle, Speed Air Man, Spellcast, Spiral Wave, SpookyMulder, SpuriousQ, Squigish, Sry85, Stan Shebs, Starexplorer, Statalyzer, Stattouk, Stefan, Stellmach, StephenDawson, Stephenb, SteveMcCluskey, Steven J. Anderson, Steveo2, StoptheDatabaseState, Stwalkerster, Summabrian, Sundar, Supabanakid, SuperExpress, Superm401, Supermat, Supernova0, Surachit, Svdragunov, Sweetness46, Swpb, Symo85, Syndicate, Szajci, Tails Soda, Tamfang, Tangotango, Tanzania, Tariqabjotu, Tarotcards, Tarquin, Tarquin Binary, Tastycheeze, Tbayboy, Tdadamemd, Tdvance, TeaDrinker, Template namespace initialisation script, Terminal157, TexasAndroid, Th1rt3en, The Enlightened, The Gaon, The High Fin Sperm Whale, The Infidel, The Merciful, The Phoenix, The Placebo Effect, The Rambling Man, The Singing Badger, The Thing That Should Not Be, The Tom, The shaggy one, TheCodeman4, TheTruth2, Theda, Thedjatclubrock, Thierry Caro, Thomasedavis, Thorpe, Thunderbrand, Thw1309, Tianxiaozhang, Tideflat, Tigah, TimShell, Time3000, Timir2, Timwi, Tinctorius, Tipou, Titoxd, Tkbwik, Tktktk, Tlogmer, Tom, Tommi Ronkainen, Tompw, Tomruen, Tonic123, Tony1, Topbanana, Tothebarricades.tk, Tpbradbury, TransistorsGoneWild, Traroth, Travelbird, Treybien, Tricky Victoria, Trottaman471, Trunks6, Trusilver, Tslocum, Ttsenis, Tulkolahten, Turnteeth, Turok 11, Turok 12, Turok 16, Turok 5, TutterMouse, Tuvas, TxMCJ, Tycho Magnetic Anomaly-1, Tyler148, Tyr Anasazi, U.S.A.U.S.A.U.S.A., UberScienceNerd, Umlautbob, UncleBubba, Uniqueuponhim, Urhixidur, Us441, Ushionna, Vary, Vasi, Vicki Rosenzweig, Viridian, VoidLurker, Voldemort, Voortle, Vowlesy, Voxii, Vrenator, Vsmith, Vy0123, Vyznev Xnebara, Watch37264, Wavelength, Wayward, Wdfarmer, Weatherman1126, Wenli, Wenny888, Werdan7, Werothegreat, Weyes, Whatismetric, Whcernan, Who, Whoop whoop pull up, WikHead, Wiki alf, Wikianon, Wikicheng, Wikiwatchers, Will.i.am, WillMak050389, Wilsjrwinners, Wimt, WingZero, Wink wink, Wise1990, Wisq, Wknight94, Wmahan, Wolfkeeper, WolfmanSF, Worldtraveller, XJamRastafire, Xammer, Xanzzibar, Xenonice, Xession, Xiahou, Yamamoto Ichiro, Yelyos, Yintan, Yodaroxmysox, Yupik, Yuya takeuchi, Zafiroblue05, Zanaq, Zaphodthewise, Zbayz, Zondor, Zowie, Zsinj, Zvika, Zzuuzz, Zzyzx11, Ævar Arnfjörð Bjarmason, Τις, Товарищ, یرون س ارای, 2399 anonymous edits

Near-Earth_object *Source*: http://en.wikipedia.org/w/index.php?title=Near-Earth_object *Contributors*: A bit iffy, Amberroom, Amrix, Anetode, Antonsusi, Anyeverybody, Arcayne, AstaG, Astral, Ataleh, Awickert, Baccyak4H, Bachrach44, Beaucouplusneutre, Beland, Benbest, Bicanashy, Bporopat, Branlon, Breno, Bryan Derksen, Bubba73, Carcharoth, Cassini83, CesarB, Chesnok, Chrismichener1, CommonsDelinker, Conversion script, Craigsjones, Cratylus3, Css, Curps, Dark Tea, Darrenlatson, Darth Panda, Davidcannon, Dawnseeker2000, Denisarona, Dorftrottel, Dpenn89, Dwlegg, Elonka, Epbr123, Eric Kvaalen, Erisie, Euyyn, Everything counts, Fartherred, Fjörgynn, Fotaun, FrankTobia, Gaius Cornelius, Gene Nygaard, Gioto, Glenn, Gravitophoton, Green caterpillar, Greenrd, GregorB, Ground Zero, Haim Berman, Hans Dunkelberg, Harryzilber, HopeSeekr of xMule, Hu, IanManka, Icairns, Ikluft, Imaginaryoctopus, Iohannes Animosus, Ixfd64, JDG, JIP, Jeandré du Toit, Jeepo, John Belushi, JorisvS, Joseph Solis in Australia, Kafziel, Kagemaru16, Kheider, Kinhull, Kwamikagami, LanceBarber, LarryGilbert, Larry_Sanger, Le sacre, Lionboy-Renae, Looxix, Lowe4091, Lunokhod, MATRIX, Manchurian candidate, Martijn Hoekstra, Matubo, Mav, Mdy66, Mets501, Michael Hardy, MikeCapone, Mikenorton, Mintleaf, Mitch Ames, Modest Genius, Mosesofmason, Mr Stephen, Mu301, Murgh, Muro de Aguas, Mzmadmike, N2e, NJR ZA, Neutrality, Ng.j, Nighend, Nr5278, Nurg, Oem2s3, Onsly, Oonissie, Originalwana, PFSLAKES1, Patmic2502, Patrick, PedroPVZ, Perry Middlemiss, Petersam, Phaldo, Phyrlight, Pohick2, Poodleboy, Primium mobile, PsuedoName, PuzzletChung, Q43, Qshio, RSStockdale, RedSpruce, Reyk, Rich Farmbrough, Rmhermen, Rockfang, Rod57, Roentgenium111, Rothorpe, RoyBoy, Rschweickart, Rst20xx, Ruslik0, Ryanjo, SAE1962, SMcCandlish, Sailbad, Saros136, Savidan, Science enthusiast343, Scotthatton, Sdsds, Serendipodous, Silence, Simulation18, Solipsist, Sonicology, Spark010, Srschu273, Srwalden, SunSw0rd, TUF-KAT, Tbhotch, Terraflorin, Th1rt3en, The Anome, Thehotelambush, Trafford09, Trilobitealive, Twas Now, UltimateDarkloid, Under22Entreprenuer, Urhixidur, Uvaduck, Uvaphdman, Wwheaton, Xenobiologista, Yankees76, Zafiroblue05, Zaqarbal, Zeimusu, 159 anonymous edits

Centaur_(minor_planet) *Source*: http://en.wikipedia.org/w/index.php?title=Centaur_%28minor_planet%29 *Contributors*: (, Acalamari, Alansohn, Argo Navis, Avinash.royyuru, Backin72, Beetstra, Bryan Derksen, CFLeon, Chesnok, Ckatz, CuriousEric, Curps, David Eppstein, Deuar, Dread83, Duoduoduo, Egil, Eurocommuter, Fjörgynn, Fotaun, Gene Nygaard, Geoffrey.landis, Giftlite, Gurch, Gveret Tered, Hike395, Icairns, Ilmari Karonen, Iridia, John Hyams, JorisvS, Joseph Solis in Australia, Jscotti, Kbdank71, Kheider, Knowledge Seeker, Kwamikagami, Ling.Nut, Lowe4091, Maxim Razin, Mejor Los Indios, Mgmirkin, Michaelmas1957, Muma, Murgh, Njardarlogar, Novangelis, PamD, PedroPVZ, Piledhigheranddeeper, Pt, Rich Farmbrough, Rjwilmsi,

Rmhermen, Roentgenium111, Ruslik0, Samdacruel, Sardanaphalus, Satori, Serendipodous, Serpentinite, Sethhater123, Spartan, Tamfang, Template namespace initialisation script, The Singing Badger, The Tom, The shaggy one, Thebiggnome, Tom harrison, Tothebarricades.tk, Trafford09, Ufviper, Ulflarsen, Urhixidur, WolfmanSF, Δ, عباد مجاهد ديرانية, 41 anonymous edits

Trojan_(astronomy) *Source*: http://en.wikipedia.org/w/index.php?title=Trojan_%28astronomy%29 *Contributors*: AdnanSa, All Is One, Amakuha, AndyKali, Baseballrocks538, BatteryIncluded, BilCat, Bryan Derksen, CSZero, Christopher Cooper, Creol, Curtis Clark, Dawnseeker2000, DerBorg, Drbogdan, Enchanter, Fjörgynn, Frankie1969, Frecklefoot, GTBacchus, Goustien, Grósznyó, Hairy Dude, Hans Dunkelberg, Hellbus, Jazzlvraz, JorisvS, Kevin Myers, Kjkolb, Kwamikagami, Lavallen, Leslie Mateus, Loadmaster, Mathias-S, Maurice Carbonaro, Medeis, Michael Hardy, Mimihitam, Mitch Ames, Njardarlogar, Ohms law, Piledhigheranddeeper, Qurq, RJHall, Rgwc, Rjwilmsi, Roentgenium111, Rothorpe, Ruslik0, Serasuna, Serendipodous, Skizzik, Spacepotato, Tamfang, Tbhotch, The Singing Badger, The Tom, TheStarmon, Traxs7, Urhixidur, Westley Turner, Wkharrisjr, WolfmanSF, XavierGreen, Yworo, 34 anonymous edits

Minor_planet *Source*: http://en.wikipedia.org/w/index.php?title=Minor_planet *Contributors*: 3coma14, Aaron of Mpls, Aesopos, Alfons2, An Justified Wikipedian, Arsia Mons, Asikal1, Bryan Derksen, Chesnok, Ckatz, David Shepheard, Dicklyon, Eekster, Fartherred, Fjörgynn, Glenn L, Gpermant, GraL, Interstellar Man, JHunterJ, Jolielegal, JorisvS, Kayau, Kheider, Kwamikagami, LeaveSleaves, MER-C, Max rspct, Merovingian, Michael Hardy, Miremare, Murgh, Nergaal, Novangelis, Nubiatech, RJHall, Raymondwinn, Reyk, Rmhermen, Roentgenium111, Rorschach, Rothorpe, Ruslik0, Serendipodous, Sfahey, Shirt58, Sobreira, Soulkeeper, Spacepotato, Spidern, The Tom, The way, the truth, and the light, Thingg, Tlesher, Trekphiler, Ultra two, Urhixidur, Vaĝinator, Velpaedia Jenkuklordanus, Wikiprodder, Wiklol, William Avery, Willking1979, Woohookitty, WooperJeff, Wwannsda, 74 anonymous edits

Kuiper_belt *Source*: http://en.wikipedia.org/w/index.php?title=Kuiper_belt *Contributors*: @pple, A2-computist, Aajacksoniv, Aaron of Mpls, Abbalkea2010, Acer, Acom, AdSR, Aditya, Alfio, Altenmann, Amaurea, Andre Engels, Andres, Arctic.gnome, ArgGeo, Art LaPella, AshLin, Astor14, AstroHurricane001, AstroMark, AstroNomer, Ataleh, Audriusa, Av0id3r, Avsa, Barneca, Bassem18, Bender235, Berek, BertSen, Bfigura's puppy, Bigger digger, BlueMoonlet, Bobo192, Bolo1729, Bongwarrior, Borgx, Boston, Brian0918, Brighterorange, Brim, Brion VIBBER, Bŕyan Derksen, Bubba73, C777, CWenger, Caerwine, Cam, Carbuncle, Card, Chesnok, Christopher Parham, Closedmouth, Co149, Cobblet, Cometstyles, Conorobradaigh, Conversion script, Cookiehead, Craig Pemberton, Ctachme, CuriousEric, Curps, Cwolfsheep, Cyberia23, Cyp, Cyrius, DNewhall, Dan Austin, Danny, Deathphoenix, Deflective, Denelson83, Desibites, Deuar, Dhum Dhum, Downhighest, Drvancampen, Ec5618, Ed Cormany, Egil, Eleassar777, ElectricValkyrie, Ellywa, Elockard, Eob, Epbr123, Eric Wester, Ericg, Escape Orbit, Eurocommuter, Evercat, Evil Monkey, Exir Kamalabadi, FF2010, Fenneth, Firsfron, Fivemack, Fjörgynn, Fluorhydric, Foosher, Fredrik, GabrielEvans, Garglebutt, Giftlite, Gilliam, Glenn, Gnickett1, GoingBatty, Graham87, Grtamlinb, Guessing Game, Hairy Dude, Harold f, Harp, HarryAlffa, Harryboyles, Headbomb, Heron, Hike395, Howcheng, HumphreyW, Iamunknown, Ignoranteconomist, Ilmari Karonen, Imgayftwyes, Inge-Lyubov, Iridescent, Iridia, Ishel99, J P, JDT1991, JMK, JamesFox, Jan.Kamenicek, Jann, Jarry1250, JasonAQuest, Jasongetsdown, Jay Litman, Jeandré du Toit, Jeffmedkeff, Jeronimo, Jk2693, Joedeshon, Joelwest, John, John Belushi, John Hyams, John85, Jolielegal, Jon.prasad, Jonshill, Jor, JorisvS, Josh Grosse, Josiah Rowe, Jyril, KBi, Keflavich, Kevyn, Kheider, Kitsunegami, Kiviuq, Kozuch, Kwamikagami, Latulla, Lestatdelc, Lightblade, Ling.Nut, Logan, LonelyMarble, LonelyPker, Looxix, Lowe4091, Luk, Lwalt, MK, MK8, MONGO, Madkayaker, Malhonen, Martin56, Mav, Mbell, Mdotley, Mendaliv, Mentifisto, Michael Devore, Michael Hardy, Mike s, Mikegrant, Minesweeper, Monz, Mooquackwooftweetmeow, Moritheil, Mrwuggs, Murgh, Murlough23, Myrrhlin, Mzajac, Nascarfan1,000,000,000, NeilTarrant, Nemo1986, Nemu, Nephelin, Nickshanks, Nixer, Njardarlogar, Ocee, Odros, Ohconfucius, Olivier, Omnedon, OrgasGirl, Originalwana, Ottawa4ever, Outriggr, Ozone77, Pablo-flores, Palica, Peashy, Pharos, Philip Trueman, Philrosenberg, Piast93, Pizza Puzzle, Pmokeefe, Ponder, Poppy, Princejack12, RA0808, RMHED, RP88, RadicalOne, Rasimpson, Rathanor, Rebrane, RedBLACKandBURN, RexNL, Rhodekyll, Rhort, Rich Farmbrough, Richard B, Richhoncho, Rickyrab, Rjwilmsi, Robert Merkel, RobertG, Roentgenium111, Ronhjones, RoosMargot, RossPatterson, Rothorpe, Roudy66, RoyBoy, Runningonbrains, Rursus, Ruslik0, Ruy Pugliesi, Sage of Ice, Sardanaphalus, Sardur, Saros136, Schneelocke, SchuminWeb, Scjessey, Scog, ScottyBoy900Q, Semperf, Serendipodous, Sffcorgi, Shadowjams, Shadowmorph, Shawn81, Shirulashem, Shsilver, Sich Bojan, Sideways713, Sietse Snel, Sintaku, Sir Nicholas de Mimsy-Porpington, Smartech, Some jerk on the Internet, Something14, Spellmaster, StephenBuxton, Stevenj, Stirling Newberry, Sturmde, T-borg, TTE, Tamfang, Tarotcards, Tedzsee, Tegles, Template namespace initialisation script, TeunSpaans, The Other Saluton, The Thing That Should Not Be, The Tom, The Yeti, TheMadBaron, TheMightyOrb, Tide rolls, TimeCruiserMike, Timwi, Tom, TomXP411, Tomchiukc, Treisijs, Triplepickle815, Triwbe, Tronno, Tuvas, Tycho Magnetic Anomaly-1, Ufviper, Uncle G, Upquark, Urhixidur, Vahagn Petrosyan, Vanis314, Vegeta624, Vicki Rosenzweig, Vimalkalyan, Vuong Ngan Ha, Wayne Hardman, Wiki alf, Wikipelli, WilyD, WolfmanSF, Writtenonsand, Wysprgr2005, XJamRastafire, XLerate, Xiahou, Ylee, Yosri, Zero sharp, Zhatt, Саша Стефановић, $_0$x, 413 anonymous edits

Image Sources, Licenses and Contributors

Image:4 Vesta 1 Ceres Moon at 20 km per px.png *Source*: http://en.wikipedia.org/w/index.php?title=File:4_Vesta_1_Ceres_Moon_at_20_km_per_px.png *License*: unknown *Contributors*: NASA, composition by Urhixidur

Image:Ceres symbol.svg *Source*: http://en.wikipedia.org/w/index.php?title=File:Ceres_symbol.svg *License*: unknown *Contributors*: Aquantrum, Crei0, Galphi, Lexicon, Ramiy, Ruslik0, Sarang, Sarilho1, Urhixidur, 4 anonymous edits

Image:Ceres2.svg *Source*: http://en.wikipedia.org/w/index.php?title=File:Ceres2.svg *License*: unknown *Contributors*: Varun Kumar

Image:Ceres3.svg *Source*: http://en.wikipedia.org/w/index.php?title=File:Ceres3.svg *License*: unknown *Contributors*: Varun Kumar

Image:2Pallas symbol.svg *Source*: http://en.wikipedia.org/w/index.php?title=File:2Pallas_symbol.svg *License*: unknown *Contributors*: User:KleinKlio

Image:2 Pallas.svg *Source*: http://en.wikipedia.org/w/index.php?title=File:2_Pallas.svg *License*: unknown *Contributors*: Vectorized by

Image:Juno symbol.svg *Source*: http://en.wikipedia.org/w/index.php?title=File:Juno_symbol.svg *License*: unknown *Contributors*: Lexicon, Sarang, Urhixidur

Image:3 Juno (1).png *Source*: http://en.wikipedia.org/w/index.php?title=File:3_Juno_(1).png *License*: unknown *Contributors*: User:Urhixidur

Image:Symbol 3.jpg *Source*: http://en.wikipedia.org/w/index.php?title=File:Symbol_3.jpg *License*: unknown *Contributors*: Bkell, Garion96, Pickom, RP88

Image:Vesta symbol.svg *Source*: http://en.wikipedia.org/w/index.php?title=File:Vesta_symbol.svg *License*: unknown *Contributors*: G.dallorto, Lexicon, Rmhermen, Sarang, Urhixidur

Image:4 Vesta (0).svg *Source*: http://en.wikipedia.org/w/index.php?title=File:4_Vesta_(0).svg *License*: unknown *Contributors*: Roland Geider

Image:Simbolo di Vesta.svg *Source*: http://en.wikipedia.org/w/index.php?title=File:Simbolo_di_Vesta.svg *License*: unknown *Contributors*: Original uploader was Kwamikagami at en.wikipedia

Image:4 Vesta Unsimplified Symbol.svg *Source*: http://en.wikipedia.org/w/index.php?title=File:4_Vesta_Unsimplified_Symbol.svg *License*: unknown *Contributors*: Original uploader was Tooto at en.wikipedia

File:5 Astraea symbol alternate.svg *Source*: http://en.wikipedia.org/w/index.php?title=File:5_Astraea_symbol_alternate.svg *License*: unknown *Contributors*: Wilson, John

Image:5 Astraea Symbol.svg *Source*: http://en.wikipedia.org/w/index.php?title=File:5_Astraea_Symbol.svg *License*: unknown *Contributors*: User:Rmhermen

Image:6 Hebe Astronomical Symbol.svg *Source*: http://en.wikipedia.org/w/index.php?title=File:6_Hebe_Astronomical_Symbol.svg *License*: unknown *Contributors*: User:Rmhermen

Image:7 Iris Astronomical Symbol.svg *Source*: http://en.wikipedia.org/w/index.php?title=File:7_Iris_Astronomical_Symbol.svg *License*: unknown *Contributors*: Editor at Large, Urhixidur, Wikiborg

Image:8 Flora Astronomical Symbol.svg *Source*: http://en.wikipedia.org/w/index.php?title=File:8_Flora_Astronomical_Symbol.svg *License*: unknown *Contributors*: Editor at Large, Urhixidur

Image:9 Metis symbol.svg *Source*: http://en.wikipedia.org/w/index.php?title=File:9_Metis_symbol.svg *License*: unknown *Contributors*: User:Bryan Derksen

File:10 Hygeia symbol alternate.svg *Source*: http://en.wikipedia.org/w/index.php?title=File:10_Hygeia_symbol_alternate.svg *License*: unknown *Contributors*: Hind, J. R.

Image:10 Hygiea Astronomical Symbol.svg *Source*: http://en.wikipedia.org/w/index.php?title=File:10_Hygiea_Astronomical_Symbol.svg *License*: unknown *Contributors*: User:Editor at Large

File:11 Parthenope symbol alternate.svg *Source*: http://en.wikipedia.org/w/index.php?title=File:11_Parthenope_symbol_alternate.svg *License*: unknown *Contributors*: Wilson, John

Image:11 Parthenope symbol.svg *Source*: http://en.wikipedia.org/w/index.php?title=File:11_Parthenope_symbol.svg *License*: unknown *Contributors*: User:Editor at Large

Image:12 Victoria symbol.svg *Source*: http://en.wikipedia.org/w/index.php?title=File:12_Victoria_symbol.svg *License*: unknown *Contributors*: User:Editor at Large

File:13 Egeria symbol.svg *Source*: http://en.wikipedia.org/w/index.php?title=File:13_Egeria_symbol.svg *License*: unknown *Contributors*: Hind, John Russell

File:14 Irene symbol.svg *Source*: http://en.wikipedia.org/w/index.php?title=File:14_Irene_symbol.svg *License*: unknown *Contributors*: John Wilson

Image:15 Eunomia symbol.svg *Source*: http://en.wikipedia.org/w/index.php?title=File:15_Eunomia_symbol.svg *License*: unknown *Contributors*: User:Bryan Derksen

File:16 Psyche symbol.svg *Source*: http://en.wikipedia.org/w/index.php?title=File:16_Psyche_symbol.svg *License*: unknown *Contributors*: J. R. Hind

Image:17 Thetis symbol.png *Source*: http://en.wikipedia.org/w/index.php?title=File:17_Thetis_symbol.png *License*: unknown *Contributors*: Symbol created by R. Luther

Image:18 Melpomene symbol.svg *Source*: http://en.wikipedia.org/w/index.php?title=File:18_Melpomene_symbol.svg *License*: unknown *Contributors*: Hind, John Russell

File:19 Fortuna symbol.svg *Source*: http://en.wikipedia.org/w/index.php?title=File:19_Fortuna_symbol.svg *License*: unknown *Contributors*: Hind, John Russell

Image:26 Proserpina symbol.svg *Source*: http://en.wikipedia.org/w/index.php?title=File:26_Proserpina_symbol.svg *License*: unknown *Contributors*: Symbol created by R. Luther

Image:28 Bellona symbol.svg *Source*: http://en.wikipedia.org/w/index.php?title=File:28_Bellona_symbol.svg *License*: unknown *Contributors*: User:Editor at Large

Image:29 Amphitrite symbol.svg *Source*: http://en.wikipedia.org/w/index.php?title=File:29_Amphitrite_symbol.svg *License*: unknown *Contributors*: Chauncey A. Goodrich and Noah Porter, editors

Image:35 Leukothea symbol.png *Source*: http://en.wikipedia.org/w/index.php?title=File:35_Leukothea_symbol.png *License*: unknown *Contributors*: G. Rumker and C.A.F. Peters

Image:37 Fides symbol.svg *Source*: http://en.wikipedia.org/w/index.php?title=File:37_Fides_symbol.svg *License*: unknown *Contributors*: R. Luther

Image:243 ida.jpg *Source*: http://en.wikipedia.org/w/index.php?title=File:243_ida.jpg *License*: unknown *Contributors*: NASA/JPL

Image:Asteroid 2004 FH.gif *Source*: http://en.wikipedia.org/w/index.php?title=File:Asteroid_2004_FH.gif *License*: unknown *Contributors*: ComputerHotline, Danim, Mattes, Njaelkies Lea, Quasipalm, Romanm, Tungsten, Vesta, West, Yann, Yarl

Image:InnerSolarSystem-en.png *Source*: http://en.wikipedia.org/w/index.php?title=File:InnerSolarSystem-en.png *License*: unknown *Contributors*: Original uploader was Mdf at en.wikipedia

Image:Moon and Asteroids 1 to 10.svg *Source*: http://en.wikipedia.org/w/index.php?title=File:Moon_and_Asteroids_1_to_10.svg *License*: unknown *Contributors*: User:Vystrix Nexoth

Image:Ceres optimized.jpg *Source*: http://en.wikipedia.org/w/index.php?title=File:Ceres_optimized.jpg *License*: unknown *Contributors*: NASA, ESA, J. Parker (Southwest Research Institute), P. Thomas (Cornell University), and L. McFadden (University of Maryland, College Park)

image:masses of asteroids vs main belt.png *Source*: http://en.wikipedia.org/w/index.php?title=File:Masses_of_asteroids_vs_main_belt.png *License*: unknown *Contributors*: User:Kwamikagami

Image:(253) mathilde.jpg *Source*: http://en.wikipedia.org/w/index.php?title=File:(253)_mathilde.jpg *License*: unknown *Contributors*: NASA

Image:433eros.jpg *Source*: http://en.wikipedia.org/w/index.php?title=File:433eros.jpg *License*: unknown *Contributors*: NASA/NEAR Project (JHU/APL).

Image:951 Gaspra.jpg *Source*: http://en.wikipedia.org/w/index.php?title=File:951_Gaspra.jpg *License*: unknown *Contributors*: Alaniaris, BokicaK, Papa November, Ruslik0, The Evil IP address, Tryphon, Yonidebest, 5 anonymous edits

File:Dawn-image-070911.jpg *Source*: http://en.wikipedia.org/w/index.php?title=File:Dawn-image-070911.jpg *License*: unknown *Contributors*: NASA/JPL-Caltech/UCLA/MPS/DLR/IDA

Image:Giuseppe Piazzi.jpg *Source*: http://en.wikipedia.org/w/index.php?title=File:Giuseppe_Piazzi.jpg *License*: unknown *Contributors*: Howcheng, Lalupa, Lotse, Papa November, Rentier, Rootology, Ruslik0

Image:Main belt i vs a.png *Source*: http://en.wikipedia.org/w/index.php?title=File:Main_belt_i_vs_a.png *License*: unknown *Contributors*: Piotr Deuar

Image:AllendeMeteorite.jpg *Source*: http://en.wikipedia.org/w/index.php?title=File:AllendeMeteorite.jpg *License*: unknown *Contributors*: User:Vesta

Image:Main belt e vs a.png *Source*: http://en.wikipedia.org/w/index.php?title=File:Main_belt_e_vs_a.png *License*: unknown *Contributors*: Original uploader was Deuar at en.wikipedia

Image:Kirkwood Gaps.svg *Source*: http://en.wikipedia.org/w/index.php?title=File:Kirkwood_Gaps.svg *License*: unknown *Contributors*: based on plot by Alan Chamberlain, JPL/Caltech

Image:zodiacal.jpg *Source*: http://en.wikipedia.org/w/index.php?title=File:Zodiacal.jpg *License*: unknown *Contributors*: Original uploader was Serendipodous at en.wikipedia

Image:Asteroid proper elements i vs e.png *Source*: http://en.wikipedia.org/w/index.php?title=File:Asteroid_proper_elements_i_vs_e.png *License*: unknown *Contributors*: Bryan Derksen, Deuar, Papa November, Ruslik0, WinstonSmith, Yann

Image:Dawn Flight Configuration 2.jpg *Source*: http://en.wikipedia.org/w/index.php?title=File:Dawn_Flight_Configuration_2.jpg *License*: unknown *Contributors*: GDK

Image:AsteroidIncAu.png *Source*: http://en.wikipedia.org/w/index.php?title=File:AsteroidIncAu.png *License*: unknown *Contributors*: Original uploader was Dreg743 at en.wikipedia

Image:Asteroid osculating vs proper elements.png *Source*: http://en.wikipedia.org/w/index.php?title=File:Asteroid_osculating_vs_proper_elements.png *License*: unknown *Contributors*: Original uploader was at zh.wikipedia

File:Solar System size to scale.svg *Source*: http://en.wikipedia.org/w/index.php?title=File:Solar_System_size_to_scale.svg *License*: unknown *Contributors*: User:Cmglee, User:Farry

File:Solarsystem3DJupiter.gif *Source*: http://en.wikipedia.org/w/index.php?title=File:Solarsystem3DJupiter.gif *License*: unknown *Contributors*: User:Lookang

File:Oort cloud Sedna orbit.svg *Source*: http://en.wikipedia.org/w/index.php?title=File:Oort_cloud_Sedna_orbit.svg *License*: unknown *Contributors*: User:Holek

File:Venustransit 2004-06-08 07-49.jpg *Source*: http://en.wikipedia.org/w/index.php?title=File:Venustransit_2004-06-08_07-49.jpg *License*: unknown *Contributors*: Ahoerstemeier, Avian, ComputerHotline, Crux, Dbenzhuser, Duckysmokton, Enirac Sum, J 1982, J.delanoy, LA2, Martwinn, Rat at WikiFur, Skab, 1 anonymous edits

File:Heliospheric-current-sheet.gif *Source*: http://en.wikipedia.org/w/index.php?title=File:Heliospheric-current-sheet.gif *License*: unknown *Contributors*: Werner Heil (see "other version" below).

File:Terrestrial planet size comparisons.jpg *Source*: http://en.wikipedia.org/w/index.php?title=File:Terrestrial_planet_size_comparisons.jpg *License*: unknown *Contributors*: wikipedia user Brian0918

File:InnerSolarSystem-en.png *Source*: http://en.wikipedia.org/w/index.php?title=File:InnerSolarSystem-en.png *License*: unknown *Contributors*: Original uploader was Mdf at en.wikipedia

File:Gas giants in the solar system.jpg *Source*: http://en.wikipedia.org/w/index.php?title=File:Gas_giants_in_the_solar_system.jpg *License*: unknown *Contributors*: NASA

File:Comet c1995o1.jpg *Source*: http://en.wikipedia.org/w/index.php?title=File:Comet_c1995o1.jpg *License*: unknown *Contributors*: User:Miketsukunibito

File:Outersolarsystem objectpositions labels comp.png *Source*: http://en.wikipedia.org/w/index.php?title=File:Outersolarsystem_objectpositions_labels_comp.png *License*: unknown *Contributors*: 84user, Bassem, Kaldari, Peteforsyth, Poppy, Venkat.athma, Wikibob, WilyD, 5 anonymous edits

File:EightTNOs.jpg *Source*: http://en.wikipedia.org/w/index.php?title=File:EightTNOs.jpg *License*: unknown *Contributors*: Serendipodous

File:NewHeliopause 558121.jpg *Source*: http://en.wikipedia.org/w/index.php?title=File:NewHeliopause_558121.jpg *License*: unknown *Contributors*: Rehman

File:Kuiper oort.jpg *Source*: http://en.wikipedia.org/w/index.php?title=File:Kuiper_oort.jpg *License*: unknown *Contributors*: NASA

File:Milky Way Spiral Arm.svg *Source*: http://en.wikipedia.org/w/index.php?title=File:Milky_Way_Spiral_Arm.svg *License*: unknown *Contributors*: User:Surachit

File:Earth's Location in the Universe (JPEG).jpg *Source*: http://en.wikipedia.org/w/index.php?title=File:Earth's_Location_in_the_Universe_(JPEG).jpg *License*: unknown *Contributors*: User:Azcolvin429

File:Magnify-clip.png *Source*: http://en.wikipedia.org/w/index.php?title=File:Magnify-clip.png *License*: unknown *Contributors*: User:Erasoft24

File:Solar Life Cycle.svg *Source*: http://en.wikipedia.org/w/index.php?title=File:Solar_Life_Cycle.svg *License*: unknown *Contributors*: User:Oliverbeatson

File:TheSun.png *Source*: http://en.wikipedia.org/w/index.php?title=File:TheSun.png *License*: unknown *Contributors*: NASA

File:Jupiter on 2010-06-07 (captured by the Hubble Space Telescope).jpg *Source*: http://en.wikipedia.org/w/index.php?title=File:Jupiter_on_2010-06-07_(captured_by_the_Hubble_Space_Telescope).jpg *License*: unknown *Contributors*: NASA, ESA, M.H. Wong (University of Califoria, Berkeley), H.B. Hammel (Space Science Institute, Boulder, Colo.), A.A. Simon-Miller (Goddard Space Flight Center), and the Jupiter Impact Science Team

File:Saturn (planet) large.jpg *Source*: http://en.wikipedia.org/w/index.php?title=File:Saturn_(planet)_large.jpg *License*: unknown *Contributors*: Voyager 2

File:Uranus2.jpg *Source*: http://en.wikipedia.org/w/index.php?title=File:Uranus2.jpg *License*: unknown *Contributors*: NASA/JPL/Voyager mission Original uploader was Serendipodous at en.wikipedia

File:Neptune.jpg *Source*: http://en.wikipedia.org/w/index.php?title=File:Neptune.jpg *License*: unknown *Contributors*: .

File:The Earth seen from Apollo 17.jpg *Source*: http://en.wikipedia.org/w/index.php?title=File:The_Earth_seen_from_Apollo_17.jpg *License*: unknown *Contributors*: NASA. Photo taken by either Harrison Schmitt or Ron Evans (of the Apollo 17 crew).

File:Venus-real.jpg *Source*: http://en.wikipedia.org/w/index.php?title=File:Venus-real.jpg *License*: unknown *Contributors*: NASA/Ricardo Nunes

File:Mars Hubble.jpg *Source*: http://en.wikipedia.org/w/index.php?title=File:Mars_Hubble.jpg *License*: unknown *Contributors*: NASA and The Hubble Heritage Team (STScI/AURA)

File:Ganymede g1 true 2.jpg *Source*: http://en.wikipedia.org/w/index.php?title=File:Ganymede_g1_true_2.jpg *License*: unknown *Contributors*: Original uploader was GeneralPatton at en.wikipedia

File:Two Halves of Titan.png *Source*: http://en.wikipedia.org/w/index.php?title=File:Two_Halves_of_Titan.png *License*: unknown *Contributors*: NASA/JPL/Space Science Institute

File:Mercury in color - Prockter07 centered.jpg *Source*: http://en.wikipedia.org/w/index.php?title=File:Mercury_in_color_-_Prockter07_centered.jpg *License*: unknown *Contributors*: User:Papa Lima Whiskey

File:Callisto.jpg *Source*: http://en.wikipedia.org/w/index.php?title=File:Callisto.jpg *License*: unknown *Contributors*: Bricktop, Conscious, Dbenbenn, J 1982, JenVan, Kristaga, Li-sung, Ruslik0, 1 anonymous edits

File:Io highest resolution true color.jpg *Source*: http://en.wikipedia.org/w/index.php?title=File:Io_highest_resolution_true_color.jpg *License*: unknown *Contributors*: Bricktop, David Levy, Jed, Julia W, Kriplozoik, Pfctdayelise, Ruslik0, Uwe W., WikipediaMaster

File:Full Moon Luc Viatour.jpg *Source*: http://en.wikipedia.org/w/index.php?title=File:Full_Moon_Luc_Viatour.jpg *License*: unknown *Contributors*: user:Lviatour

File:Europa-moon.jpg *Source*: http://en.wikipedia.org/w/index.php?title=File:Europa-moon.jpg *License*: unknown *Contributors*: Dbenbenn, Li-sung, Ruslik0, Sobi3ch, 2 anonymous edits

File:Triton moon mosaic Voyager 2 (large).jpg *Source*: http://en.wikipedia.org/w/index.php?title=File:Triton_moon_mosaic_Voyager_2_(large).jpg *License*: unknown *Contributors*: NASA / Jet Propulsion Lab / U.S. Geological Survey

File:Titania (moon) color cropped.jpg *Source*: http://en.wikipedia.org/w/index.php?title=File:Titania_(moon)_color_cropped.jpg *License*: unknown *Contributors*: User:Ruslik0

File:PIA07763 Rhea full globe5.jpg *Source*: http://en.wikipedia.org/w/index.php?title=File:PIA07763_Rhea_full_globe5.jpg *License*: unknown *Contributors*: NASA/JPL/Space Science Institute

File:Voyager 2 picture of Oberon.jpg *Source*: http://en.wikipedia.org/w/index.php?title=File:Voyager_2_picture_of_Oberon.jpg *License*: unknown *Contributors*: NASA Original uploader was Looxix at en.wikipedia

File:Iapetus as seen by the Cassini probe - 20071008.jpg *Source*: http://en.wikipedia.org/w/index.php?title=File:Iapetus_as_seen_by_the_Cassini_probe_-_20071008.jpg *License*: unknown *Contributors*: NASA/JPL/Space Science Institute

File:Umbriel moon 1.gif *Source*: http://en.wikipedia.org/w/index.php?title=File:Umbriel_moon_1.gif *License*: unknown *Contributors*: NASA (Voyager)

File:Ariel-NASA.jpg *Source*: http://en.wikipedia.org/w/index.php?title=File:Ariel-NASA.jpg *License*: unknown *Contributors*: Original uploader was Svdmolen at nl.wikipedia

File:Dione (Mond) (30823363).jpg *Source*: http://en.wikipedia.org/w/index.php?title=File:Dione_(Mond)_(30823363).jpg *License*: unknown *Contributors*: alan taylor from Hudson, MA, usa

File:Inset-sat tethys-large.jpg *Source*: http://en.wikipedia.org/w/index.php?title=File:Inset-sat_tethys-large.jpg *License*: unknown *Contributors*: NASA / JPL

File:Enceladus from Voyager.jpg *Source*: http://en.wikipedia.org/w/index.php?title=File:Enceladus_from_Voyager.jpg *License*: unknown *Contributors*: Andro96, Bricktop, Li-sung, Ruslik0, Sobi3ch, Startaq, Uwe W.

File:Miranda.jpg *Source*: http://en.wikipedia.org/w/index.php?title=File:Miranda.jpg *License*: unknown *Contributors*: From de.wiki (NASA image)

File:Proteus (Voyager 2).jpg *Source*: http://en.wikipedia.org/w/index.php?title=File:Proteus_(Voyager_2).jpg *License*: unknown *Contributors*: Voyager 2, NASA

File:Mimas moon.jpg *Source*: http://en.wikipedia.org/w/index.php?title=File:Mimas_moon.jpg *License*: unknown *Contributors*: NASA

File:Hyperion in natural colours.jpg *Source*: http://en.wikipedia.org/w/index.php?title=File:Hyperion_in_natural_colours.jpg *License*: unknown *Contributors*: Uwe W.

File:Phoebe cassini.jpg *Source*: http://en.wikipedia.org/w/index.php?title=File:Phoebe_cassini.jpg *License*: unknown *Contributors*: NASA/JPL/Space Science Institute

File:PIA12714 Janus crop.jpg *Source*: http://en.wikipedia.org/w/index.php?title=File:PIA12714_Janus_crop.jpg *License*: unknown *Contributors*: NASA / Jet Propulsion Laboratory / Space Science Institute

File:Amalthea (moon).gif *Source*: http://en.wikipedia.org/w/index.php?title=File:Amalthea_(moon).gif *License*: unknown *Contributors*: Giorgiomonteforti, Nolanus, Quadell, Ruslik0, Uwe W.

File:PIA09813 Epimetheus S. polar region.jpg *Source*: http://en.wikipedia.org/w/index.php?title=File:PIA09813_Epimetheus_S._polar_region.jpg *License*: unknown *Contributors*: NASA/JPL/Space Science Institute

File:Prometheus 12-26-09a.jpg *Source*: http://en.wikipedia.org/w/index.php?title=File:Prometheus_12-26-09a.jpg *License*: unknown *Contributors*: NASA / JPL / Space Science Institute

Image:Toutatis.jpg *Source*: http://en.wikipedia.org/w/index.php?title=File:Toutatis.jpg *License*: unknown *Contributors*: Steve Ostro, JPL

Image:ESO-Asteroid Toutatis-phot-28c-04-normal.jpg *Source*: http://en.wikipedia.org/w/index.php?title=File:ESO-Asteroid_Toutatis-phot-28c-04-normal.jpg *License*: unknown *Contributors*: ESO

File:WISE Finds Fewer Asteroids near Earth.ogv *Source*: http://en.wikipedia.org/w/index.php?title=File:WISE_Finds_Fewer_Asteroids_near_Earth.ogv *License*: unknown *Contributors*: NASA

File:NearEarthAsteroidOrbitTypes.svg *Source*: http://en.wikipedia.org/w/index.php?title=File:NearEarthAsteroidOrbitTypes.svg *License*: unknown *Contributors*: User:Antonsusi

Image:Impact event.jpg *Source*: http://en.wikipedia.org/w/index.php?title=File:Impact_event.jpg *License*: unknown *Contributors*: Original uploader was Fredrik at en.wikipedia

File:1950da color 150.jpg *Source*: http://en.wikipedia.org/w/index.php?title=File:1950da_color_150.jpg *License*: unknown *Contributors*: ComputerHotline, Danim, PedroPVZ, Rmhermen

Image:Neo-chart.png *Source*: http://en.wikipedia.org/w/index.php?title=File:Neo-chart.png *License*: unknown *Contributors*: Alan B. Chamberlin

Image:Outersolarsystem objectpositions labels comp.png *Source*: http://en.wikipedia.org/w/index.php?title=File:Outersolarsystem_objectpositions_labels_comp.png *License*: unknown *Contributors*: 84user, Bassem, Kaldari, Peteforsyth, Poppy, Venkat.athma, Wikibob, WilyD, 5 anonymous edits

Image:TheKuiperBelt 42AU Centaurs.svg *Source*: http://en.wikipedia.org/w/index.php?title=File:TheKuiperBelt_42AU_Centaurs.svg *License*: unknown *Contributors*: User:Eurocommuter

File:AsbolA.GIF *Source*: http://en.wikipedia.org/w/index.php?title=File:AsbolA.GIF *License*: unknown *Contributors*: Aldo Vitagliano

Image:TheKuiperBelt Albedo and Color.svg *Source*: http://en.wikipedia.org/w/index.php?title=File:TheKuiperBelt_Albedo_and_Color.svg *License*: unknown *Contributors*: User:Eurocommuter

Image:Comet38P2067.gif *Source*: http://en.wikipedia.org/w/index.php?title=File:Comet38P2067.gif *License*: unknown *Contributors*: Orbit Viewer applet originally written and kindly provided by Osamu Ajiki (AstroArts), and further modified by Ron Baalke (JPL). Tweaks by Kevin Heider.

File:Lagrange very massive.svg *Source*: http://en.wikipedia.org/w/index.php?title=File:Lagrange_very_massive.svg *License*: unknown *Contributors*: EnEdC, JorisvS, Pieter Kuiper, Tano4595, 2 anonymous edits

File:Loudspeaker.svg *Source*: http://en.wikipedia.org/w/index.php?title=File:Loudspeaker.svg *License*: unknown *Contributors*: Bayo, Gmaxwell, Husky, Iamunknown, Mirithing, Myself488, Nethac DIU, Omegatron, Rocket000, The Evil IP address, Wouterhagens, 17 anonymous edits

File:GerardKuiper.jpg *Source*: http://en.wikipedia.org/w/index.php?title=File:GerardKuiper.jpg *License*: unknown *Contributors*: Dodo, Miraceti, Ruslik0, Svdmolen

File:Maunatele.jpg *Source*: http://en.wikipedia.org/w/index.php?title=File:Maunatele.jpg *License*: unknown *Contributors*: NASA

File:Lhborbits.png *Source*: http://en.wikipedia.org/w/index.php?title=File:Lhborbits.png *License*: unknown *Contributors*: Poppy, Rathorius, Stanlekub, 3 anonymous edits

File:Dust Models Paint Alien's View of Solar System.ogv *Source*: http://en.wikipedia.org/w/index.php?title=File:Dust_Models_Paint_Alien's_View_of_Solar_System.ogv *License*: unknown *Contributors*: NASA

File:TheKuiperBelt 75AU All.svg *Source*: http://en.wikipedia.org/w/index.php?title=File:TheKuiperBelt_75AU_All.svg *License*: unknown *Contributors*: User:Eurocommuter

File:TheKuiperBelt classes-en.svg *Source*: http://en.wikipedia.org/w/index.php?title=File:TheKuiperBelt_classes-en.svg *License*: unknown *Contributors*: User:Eurocommuter, User:Lilyu

File:Semimajorhistogramofkbos.svg *Source*: http://en.wikipedia.org/w/index.php?title=File:Semimajorhistogramofkbos.svg *License*: unknown *Contributors*: User:Rursus

File:2003 UB313 near-infrared spectrum.gif *Source*: http://en.wikipedia.org/w/index.php?title=File:2003_UB313_near-infrared_spectrum.gif *License*: unknown *Contributors*: Admrboltz, Adoniscik, Bassem, Bryan Derksen, Chromosome, EDUCA33E, Ewen, Jan.Kamenicek, Orionist, Poppy, Poulpy, 1 anonymous edits

File:TheKuiperBelt PowerLaw2.svg *Source*: http://en.wikipedia.org/w/index.php?title=File:TheKuiperBelt_PowerLaw2.svg *License*: unknown *Contributors*: User:Eurocommuter

File:TheKuiperBelt Projections 100AU Classical SDO.svg *Source*: http://en.wikipedia.org/w/index.php?title=File:TheKuiperBelt_Projections_100AU_Classical_SDO.svg *License*: unknown *Contributors*: User:Eurocommuter

File:New horizons Pluto.jpg *Source*: http://en.wikipedia.org/w/index.php?title=File:New_horizons_Pluto.jpg *License*: unknown *Contributors*: Bryan Derksen, Dream out loud, Kukanotas, Ruslik0, SchuminWeb, ShakataGaNai, 2 anonymous edits

File:Kuiper belt remote.jpg *Source*: http://en.wikipedia.org/w/index.php?title=File:Kuiper_belt_remote.jpg *License*: unknown *Contributors*: User:Audriusa

MIX
Papier aus verantwortungsvollen Quellen
Paper from responsible sources
FSC® C105338

Printed by Books on Demand GmbH, Norderstedt / Germany